喵了个咪啊去哪儿了？

[加]伊恩·菲利普斯 编著 夏楠 译

全国百佳图书出版单位
时代出版传媒股份有限公司
黄山书社

图示
狗
猫
鸟
牛
仓鼠
鼠
蛇
兔子
龟
雪貂

丢失宠物数量分布

目录

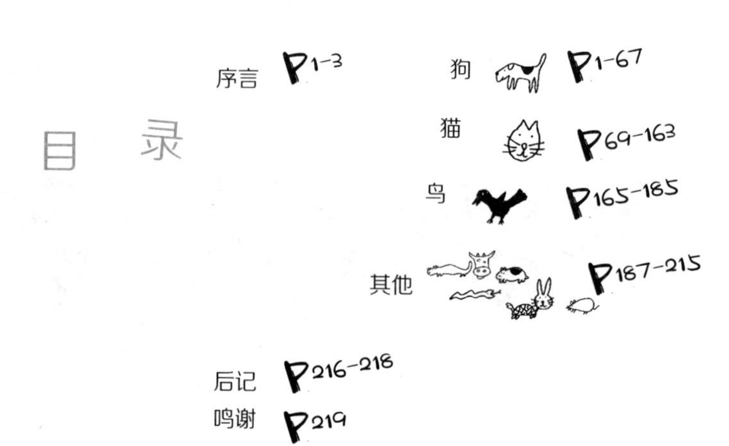

序　言

我为何如此热衷收集寻找宠物的海报？

我收集了很多寻找宠物的海报，每幅海报都是一个艺术创作，诉说了一个伤心的故事。这些故事饱含着主人对其宠物的爱、亲情和失去它们后的悲伤。尽管这些海报成本低廉且极易损坏，宠物的主人还是极其用心地将其绘制出来，并将这份爱展示给每一个站在电线杆前阅读的人。

我收集了各种类型寻找宠物的海报，从小猫小狗到白鼬公牛，无所不包。其中一张是一个女孩寻找她走丢小狗的海报，她就像呼唤失去联系的男友那样深情地呼唤她的爱犬：“一定让他打电话给我。”还有一张海报只写了一句可怜的请求，“小乌龟丢了，请帮忙找到它。”

当我还是个孩子的时候，父母只允许我养些像金鱼、仓鼠之类的小动物。小仓鼠们总是想从它们的笼子中“越狱”然后逃之夭夭，因为我总是在冰箱的下面或墙洞里找到我的小仓鼠们，所以我从未在邻里间粘贴“走丢的仓鼠”的海报。很快，我便在美丽的安大略北部长大成人了。

当我在瑞士居住的时候，我有了收集寻找宠物海报的爱好。

当时我的室友养了一只名叫“纳瓦”的猫。纳瓦总是会爬出窗户，然后沿着窄窄的窗沿去“拜访”隔壁邻居家的猫。有一天，纳瓦从我们五层高的公寓楼顶跌落后丢失了。我的室友很伤心，在社区到处贴满了寻猫海报。两三周后，我们接到了一个兽医的电话。他在电话里说他为纳瓦受伤的脚掌、腿以及爪子都做了手术，并要求我的室友必须支付他 3000 瑞士法郎的医药费才可以将纳瓦领回。我的室友按照要求把钱打给了他。几年后，纳瓦在我室友的新家中再次出走，这次它再也没有回来。

我很想知道在世界的其他地方寻找宠物的海报是什么样的，于是我在各种类型的电子杂志上登广告，并通过网络咨询家人、朋友、笔友和艺术家。后来我收集寻找宠物海报的消息传播开来，很多人将他们的海报从澳大利亚、日本、欧洲，穿越了南美和北美发到我家里。我甚至还收到了灭虫项圈、狗牌、绘有小鸡的图画和一大堆信。其中有一封信来自冰岛。信上说，由于在冰岛从未有人丢失过宠物，所以遍寻大街也找不到一张关于寻找宠物的海报。还有一封来自荷兰的信上写道：“我们从来不做张贴海报这样的事，我们丢失了宠物后要做的仅仅就是到街上再买个新的回来。”

这本书里将展示我收藏的最喜欢的海报：一张是找寻一只名叫“毒药”的小耗子；一张注着“找到者奖励 10000 美元”。其他还有：寻找在抢劫、偷车甚至地震中丢失的宠物；寻找保姆弄丢的宠物；寻找布丁、猪猪、猪肉饼、多了个脚趾的小猫、残腿的小狗；寻找玛丽猫、凯蒂猫、埃尔维斯、格瑞泽贝拉，等等。

如果你也开始收藏海报，记得在你剪下海报的位置用新的海报贴上去。如果你取下一张海报，记得做好备份再贴回去十张哦。

喵喵！

[加] 伊恩 · 菲利普斯

玛丽猫： 出自迪士尼的第 20 部经典动画电影《猫儿历险记》（*The Aristocats*），这部电影在 1970 年圣诞节前夕首度在美国剧院推出上映。片中玛丽猫高贵优雅的气质吸引了大批粉丝，而她的相关产品也成为了大家竞相收藏的对象。
凯蒂猫： 日本知名卡通人物，诞生于 1974 年。这只头系红色蝴蝶结的小白猫是有史以来最赚钱的卡通形象之一。凯蒂猫满足了人们对于童真的热望，完全依靠自身的感召力，成为 20 到 21 世纪一个长盛不衰的文化符号。
埃尔维斯： 猫王的名字，美国摇滚乐史上影响力最大的歌手，有“摇滚乐之王”的美称。
格瑞泽贝拉： 英国有史以来最成功、连续公演时间最久的音乐剧《猫》中一只非常有名的猫，被称为“魅力猫”。这部音乐剧中其他知名的猫还有“领袖猫”“富贵猫”“保姆猫”“剧院猫”“摇滚猫”“犯罪猫”“迷人猫”“英雄猫”“超人猫”“魔术猫”等。

Dogs

狗 哺乳动物，种类很多，嗅觉和听觉都很灵敏。舌长而薄，可散热，毛有黄、白、黑等色，是人类最早驯化的家畜，通常被称为“人类最忠实的朋友”。有的可以训练成警犬，有的用来帮助打猎、牧羊等。亦称“犬”。在中国文化中，狗属于十二生肖之一，在十二生肖中排名第11位。狗的寿命在12~18年之间，最长的有20年以上，不过很少见。与猫的平均寿命相近。

寻找爱犬

1996年8月21日星期三下午，

我们的小狗狗在贝尔顿和里诺街区走失了。

他体重22磅，是只白色比熊犬，

有着可爱的杏黄色耳朵，走丢时没有戴项圈。

他名叫“牛排”，

如果有谁找到或者见到他，请跟我们联系，非常感谢。

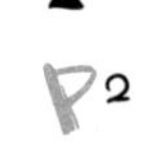

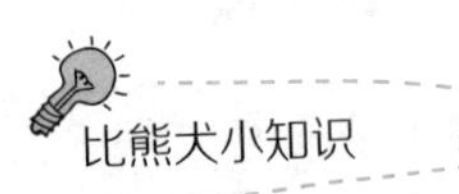

比熊犬小知识

比熊犬原产于地中海地区，是一种娇小、强健的白色粉扑型的狗，性情彬彬有礼、敏感、顽皮，活泼可爱，蓬松的小尾巴竖在后背，有着一双充满好奇的黑色眼睛。同时它的动作优雅、灵活，惹人喜爱。对主人很友善，很适合作为家养宠物。

LOST

OUR DOG WENT MISSING FROM THE BELTON STREET AND RENO STREET AREA DURING THE AFTERNOON OF WEDNESDAY, AUGUST 21, 1996. HE IS A TWENTY-TWO POUND, WHITE BICHON WITH APRICOT COLOURED EARS. HE IS NOT WEARING HIS COLLAR.

HIS NAME IS T-BONE.

IF HE IS FOUND OR IF HE HAS BEEN SEEN, PLEASE CONTACT

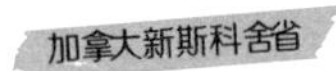

寻找“巴鲁”

我家大宝贝“巴鲁”不见了。

他体重63磅，雄性，今年11岁了，非常温顺友好。

请帮忙找找他，非常感谢。

LOST DOG
"Baru", 63 lbs, male,
11 years old, very friendly
AREA

美国夏威夷

“邓肯” 走毛了

年龄：8 岁

体重：80 磅

毛色：黄色

性格：极为温顺

定重谢！

DUNCAN IS MISSING

FRIENDLY YELLOW LAB
8YRS OLD AND 80LBS.

REWARD

寻狗启事

它是一只白色标准贵宾犬，体重8磅，

如果见到它的话请拨电话：

696-4504

→ P8

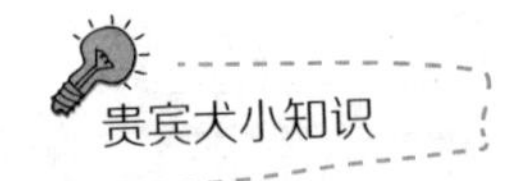

贵宾犬也称“贵妇犬”，属于非常聪明且喜欢狩猎的犬种。据猜测贵宾犬起源于德国，在那儿它以善于水中捕猎而著称。然而许多年以来，它一直被认为是法国的国犬。贵宾犬分为标准犬、迷你犬、玩具犬三种。它们之间的区别只是在于体形的大小不同。

White Standard
Poodle Puppy 8 mos.

Please Call
696-4504

我想回家！

我是一只丢了项圈的巴西狗。
我是一只披着白色大衣的绅士。
我是一只像极了狐狸的狗。

在2月13日的星期六，
我不小心走丢在山顶中心。

现今我找不到回家的路，请帮我拨打

(866) 295-8665

找到我的家人。

白色绅士需要你的帮助！

LOST DOG

lost Sat. 2/13 in the Hillcrest Medcenter area.

Tan & white Shiba Inu. Male
Name: JOEY
looks like a fox
He is missing his collar.

PLEASE HELP

Call (866) 295-8665
if you have any info

寻 找 大 型 比 格 尔 犬

我的爱犬“夏天”走丢了。

他是一只雄性比格尔犬，白、黑、茶三色，

7个月大了，性格非常好，戴着黄绿色项圈。

如果您见到了他，
请和杉山联系哦。

非常感谢您的热心帮助。

我的地址是津山市北园町3-5。

Tel:23-6306

→ P12

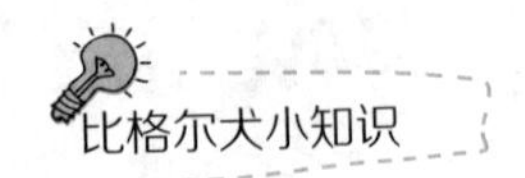

比格尔犬原产于英国，为小型猎犬，嗅觉极好，可追踪气味而猎捕胡狼、野兔、野猪、鹿等。能与其他宠物友好相处，是聪明、愉快而亲切、优秀的展示犬。我国多把其当成微型观赏犬。

さがしています。

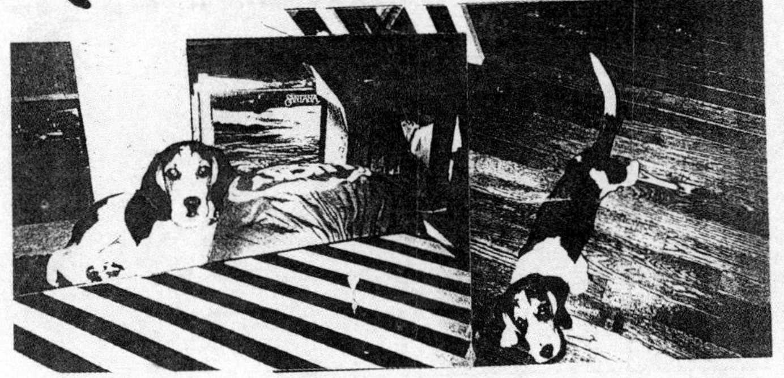

ーたずね犬ー

ビーグル犬　どんなことでもよろしいです。連らく下さい。

名前…なつ　色…白・黒・茶

生後…7ケ月　性別…メス

性格…とても元気でつかれをしらない犬です。

首輪の色、黄緑

連絡先　津山市園町3-5　杉山

23-6306

日本

寻找法国斗牛犬“比才”

（通体黑色斑纹，体重 33 磅，雄性）

我跟大卫失去了联系，如果有人知道大卫在哪儿，请让他打电话给狄安娜，电话是 (332)872-5332。

我急切地想知道我的法国斗牛犬“比才”现在在哪里。必须说明的是，它有疾病史。如果有人见到它，请联系我，谢谢。

P14

这张海报是找宠物狗的还是找男朋友的？

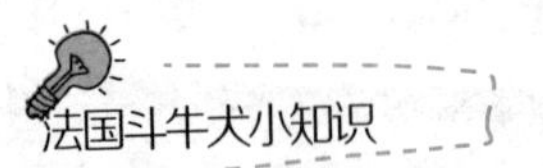

法国斗牛犬是一种活泼、聪明、敦厚、忠诚、执著、勇敢、肌肉发达的狗，骨骼沉重，被毛平滑、结构紧凑，体形中等或较小。表情显得警惕、好奇，对小孩和善，同时也是作风彪悍，能力强的优秀警卫犬。

French Bulldog named Bizet

(black brindle, 33lbs, male)

David, I lost contact. Please call Deanna at (332)872-5332. If anyone knows David, pls. tell him to call.
I just want to know how is Bizet doing. Also, I need to give you Bizet's medical history. Thank you.

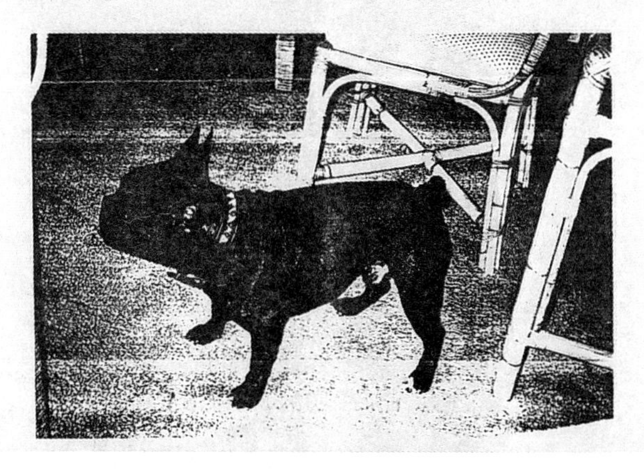

Is this a missing dog poster or a missing boyfriend?

急寻小公主“琳达”

她现年11岁，是一只巴哥犬，身子左面有个肉瘤。

请帮我找找她吧，万分感谢您的好心。

定重谢哦！

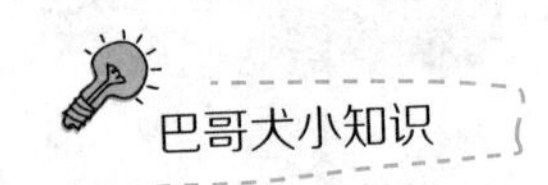

巴哥犬小知识

巴哥犬（或称哈巴狗）是一种小型短毛玩赏犬，毛质柔软而有光泽，触感顺滑，身高在25~28厘米之间，个性温驯、聪明、忠诚、友善、体贴、爱干净，寿命一般为13~14年。

REWARD

10YRS OLD "PUG"
TUMOR LF/SIDE
LINDA

美国科罗拉多州

寻 找 爱 犬

我的爱犬于1995年6月17日星期六走失。

她名叫“查理”，是只雌性西伯利亚雪橇犬，

拥有白色的尾巴！

谢谢！

西伯利亚雪橇犬小知识

西伯利亚雪橇犬又名哈士奇，是原始的古老犬种，名字的由来是源自哈士奇独特的嘶哑叫声。西伯利亚东北部的原始部落楚克奇族（Chukchi）人，用这种外形酷似狼的犬种作为最原始的交通工具来拉雪橇，并用这种狗猎取和饲养驯鹿，或者繁殖这种狗然后出售换取温饱。哈士奇性格多变，有的极端胆小，有的极端暴力，进入陆地和家庭的哈士奇都已经没有了野性，比较温顺，是一种流行于全球的宠物犬。

WANTED

E'STATO SMARRITO
SABATO 17-06-95 NEI
PRESSI DI S. ROSA, CUCCIOLO
DI CANE METICCIO, MOLTO
SOMIGLIANTE AD UN
SIBERIAN HUSKY, DI NOME
CHARLIE. SEGNI PARTICOLA-
RI: SOPRACCIGLIA E PUNTA
DELLA CODA BIANCHE.

GRAZIE

意大利

急寻“泰特莱”

重谢1000美金

爱犬“泰特莱”是一只黑白相间的雌性吉娃娃，只有四岁大。她被偷走了，如果没有饭吃她马上就会饿死的。求好心人帮忙找找她，我非常担心她！

这只狗曾上过《纽约新闻日报》的封面，标题为“无情的小偷在地铁上偷走了我的行李箱，里面有我心爱的吉娃娃宠物狗”。

吉娃娃小知识

吉娃娃（Chihuahua）属小型犬种里最小型的犬，优雅、警惕、动作迅速，以匀称的体格和娇小的体形广受人们的喜爱。有人认为此犬原产于南美，初期被印加族人视为神圣的犬种，后来传到阿斯提克族。也有人认为此犬是随西班牙的侵略者到达新世界的品种，或者在19世纪初期，从中国传入的。吉娃娃易难产，产前需尽量咨询兽医。此事性命攸关，切勿大意。

Perdido Perro

"Tetley"

DINERO PARA PERRA - $1000

La perrita de familia. Cuatro lebres y cuatro años la pura tiene . Blanco y negro chihuahua. La comida de perra esta para VIDA O MORIR

This dog appeared on the cover of "New York Newsday" with the headline: Heartless subway thief steals duffel bag bearing beloved pet chihuahua.

美国纽约

寻狗启事

宝宝在哭泣

{微型雪纳瑞犬}

我的一只雌性微型雪纳瑞犬不小心走丢了，她有着长长的尾巴。要是您发现了她，请和我联系。

Tel:784-9329

P22

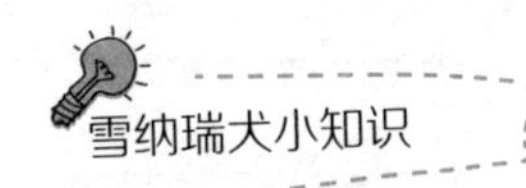

雪纳瑞犬小知识

雪纳瑞属于梗类犬的一种，源起于15世纪的德国，是梗犬类中唯一不含英国血统的品种。其名字Schnauzer是德语的“口吻”之意。它们精力充沛、活泼。雪纳瑞分为“标准雪纳瑞”、“迷你雪纳瑞”和“巨型雪纳瑞”三个品种。

LOST FEMALE
DOG
CHILDREN CRYING
(MINIATURE
SCHNAUZER)
DOES NOT HAVE
SCHNAUZER CUT. HAS
LONG TAIL. PIEASE CALL
784-9329

美国德克萨斯州

爱犬“波弟”大搜索

请打电话

967-570-1967 “波弟”走丢了！！

名字叫“波弟”，7岁，黑色毛皮，戴着粉色项圈。

如果您见到了它，请和我联系，非常感谢。

P24

LOST DOG

迷子犬捜してます!

Please Call.. お電話下さい!!

967-570-1967

Name: Boby, 7- years old

Black Pag. PINK Collar.

"黒パグ"です!!

Pink ピンク首輪

加拿大不列颠哥伦比亚省

寻找小狗！重赏！

我的小狗走失了。如果您见到了它，

请致电崔维斯，313-1735，必有重谢。

LITTLE
DOG
BIG
REWARD
CALL TRAVIS 313.1735

美国加利福尼亚
P27

全身布满斑点
毛发光滑柔软
猎犬

我的爱犬于

1996年1月31日走失了。

如果你见到它，请联系我，谢谢。

2725762

→ P28

STARRY SMOOTH HOUND
is missing:
last seen 31/1/96
please help find our
mascot call:
ssh

小狗走丢了

我的小狗名叫“斯玛特”，他是只雄性猎狐梗，栗色。
于星期天晚上 7 点在杉树种植园丢失，走丢时脖子上戴着项圈。
如果您见到了他，请拨打电话 4671573 或 4686917，定重谢！

猎狐梗小知识

猎狐梗的英文名为 Fox Terriers，是传统的英国梗类，精力充沛，不受控制，喜欢争斗。猎狐梗在猎狐类猎犬中是匀称性的典范。作为一种类似猎犬的梗类犬，腿不能太长也不能太短。它像一个聪明的猎人，能适应许多种地形，而且后背短。对这种梗而言，重量并不是一个重要指标，而形状、大小和轮廓对其工作而言，是非常重要的；如果它能奔跑、静候，追踪狐狸直到狐狸精疲力尽，那么，重个 1~2 磅将不是什么问题。

D E S A P A R E C E U

CÃO DE RAÇA "FOX-TERRIER", PÊLO CASTANHO E BRANCO, DESAPARECEU DE SUA CASA NA RUA DO PINHAL - LIVRAMENTO - S. JOÃO ESTORIL - NA TARDE DE DOMINGO DIA 7 DE MARÇO.

TEM COLEIRA COM O NOME "SMARTY" E Nº DE TELEFONE 4671573 OU 4686917.

DÃO-SE ALVISSARAS A QUEM ENCONTRAR OU INDICAR PARADEIRO.

葡萄牙

我的爱犬“赖伯”不见了

它有着黑色皮毛，没有戴项圈，也没有腿！

急需治疗！！！

如果您见到了它，请致电334-5015和汤姆·詹宁斯联系。

LOST BLACK LAB

No Collar, No Legs,

NEEDS Medicine!!!

Call 334-5015

Ask for Unca Tom Jennings

美国佛罗里达州

寻找爱犬

我的宠物狗于7月22日星期六下午3点走丢。她是一只雌性黄色贵宾犬，耳朵是茶色的，鼻子右侧3厘米处有黑色印痕。如有见到，请联系杉山方。地址：千叶。

Tel：3367-0478

鼻の右側に
2.3cmの黒い部分があります。

捜しています

7月22日土曜日午後3時頃に,
家から出たまま戻って来ません!

種類トイ・プードル
性別オス 2才
色 ベージュ 耳の所が茶系

お心当りの方は,お電話下さい

3367-0478 杉山方 千葉

日本

寻找爱犬

请大家帮帮我！

1岁大的来自鲁什尔姆的“扎克”于7月4日下午4点的时候在RUSHOLME的KWIK SAVE超市外面被偷了，他是一只有着异常大的耳朵的黑褐色猎犬，是杜宾犬和德国牧羊犬的后代。如果你看到他或是了解他在什么地方，请打电话给我。若谁能将他安全带回，将会有重赏。

STOLEN DOG

PLEASE CAN YOU HELP?

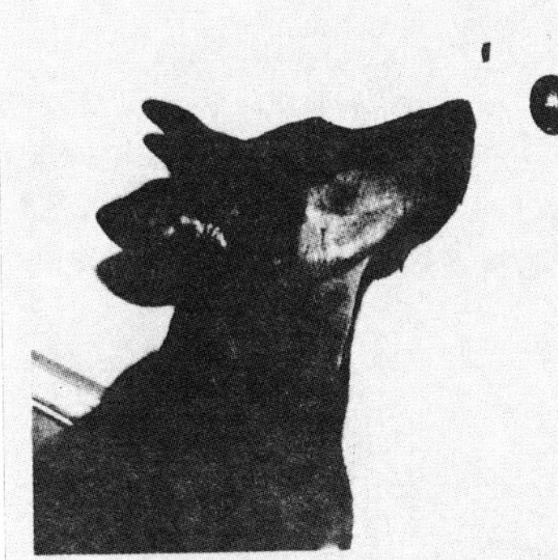

ONE YEAR OLD ZAK WAS TAKEN FROM OUTSIDE THE RUSHOLME KWIK SAVE ON THE 4th JUNE AT 4p.m. HE IS A VERY LARGE DOBERMAN/GERMAN SHEPHERD, BLACK + TAN, WITH EXCEPTIONALLY LARGE EARS. IF YOU HAVE SEEN HIM, OR HAVE ANY INFORMATION TO HIS WHEREABOUTS PLEASE RING

A REWARD WILL BE GIVEN FOR HIS SAFE RETURN.

我的爱犬“塞布丽娜”

被偷了

它是一只灰色的惠比特犬。

现特意重金悬赏来寻找它

1000 美元

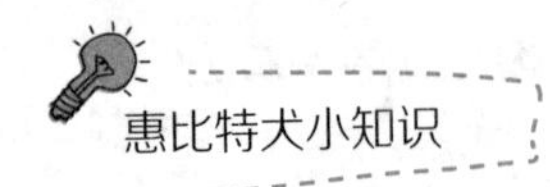

惠比特犬是真正的运动型中型猎犬，能以最少的动作跑完最长的距离。给人的印象是漂亮而和谐，肌肉发达，强壮有力，外形极度高雅优美。这种犬原产于英国，寿命一般为13~14年，四肢修长，身体瘦小柔软，是灵缇犬的缩小型，擅长狩猎兔子，又称猎兔犬。

STOLEN

~~LOST~~ WHIPPET/
GREYHOUND

"Sabrina"

STOLEN

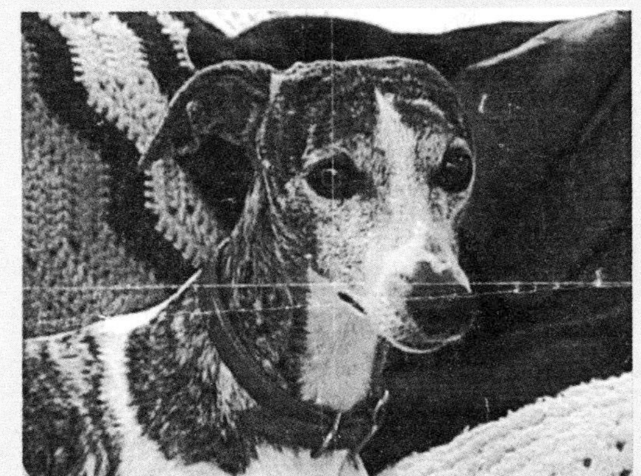

STOLEN

REWARD

$1000

美国加利福尼亚

寻 爱 犬 悬 赏 启 事

我的爱犬“波波”是一条斯塔福郡斗牛梗。

她是个姑娘，今年 2 岁了，体形又高又强壮，

脖子和下巴有白色斑纹。

她不幸于 6 月 28 日星期一走失了。

重谢 1000 美金！！！

请 大 家 帮 帮 我 ！

斯塔福郡斗牛梗小知识

斯塔福郡斗牛梗原产于英国，由斗牛犬和马士提夫犬杂交而成。雄性身高一般在35~40厘米之间，体重在12.7~17.2公斤之间；雌性身高在35~40厘米之间，体重在10.8~15.4公斤之间。斯塔福郡斗牛梗是一种短毛狗。就体形而言，其力气非常大，尽管肌肉非常发达，但其仍然十分活泼、敏捷。现代的斯塔福郡斗牛梗具有不屈不挠的品质，极度聪明和坚韧，对朋友极具感情，非常沉着，可信赖，是第一流的万能狗。

STOLEN

Monday,28th of June

Staffordshire bull terrier-FROM HIGHGATE VILLAGE

£1000 Cash reward -

For her safe return .
DESCRIPTION- Large female .
Brindle with white markings
under her neck and on her chin
2 years old.
ANSWERS TO BO-BO

PLEASE HELP.

急寻爱犬“淑女”

我将以 100 美元表达我的感激之情

基本特征：

我的爱犬“淑女”已经 3 岁了，50~60 磅重，是一只柯利牧羊犬。

她毛色金黄，腹部雪白，脚上也有很多白色的斑点。

她很喜欢人，总喜欢围着人打转。受到别人关注她会非常高兴。

她既不咬人，也不会叫，她受过很好的训练，非常听话温顺。

走丢详情：

令我非常痛心的是，“淑女”于 1992 年 6 月 28 日在洛克斯伯和央街附近走丢了。我曾经多次在她走丢的区域附近寻访，几个当地居民也非常热情地帮我提供线索，当然这期间也有人对我不理不睬。如果她是被此地的“某人”带回家了，请您务必好好照顾她，拜托了！如果您见到了她，请一定要和我联系。

P42

柯利牧羊犬小知识

柯利牧羊犬又名柯利犬、苏格兰牧羊犬，原产于英国，雄性身高一般在 61~66 厘米之间，体重在 27~34 公斤之间；雌性身高在 56~61 厘米之间，体重在 23~29 公斤之间。柯利牧羊犬是一种柔韧、结实、积极、活泼的品种，自然站立时，整齐而稳固。深且宽度适中的胸部显示出力量，倾斜的肩胛和适度弯曲的飞节显示出速度和优雅，脸部显示出非常高的智商。柯利犬给人印象深刻，是自信的化身，代表真正的和谐，每一部分都与其他部分及整体构成完美、和谐的比例。

LOST DOG
NAME "LADY"
$100.00
REWARD

FOR INFORMATION LEADING TO THE RECOVERY OF THIS OWNERS' PET.

DETAILS:

LADY WAS LOST IN THE ROXBOROUGH & YONGE ST. AREA JUNE 28/92. I HAVE SPOKEN TO SOME OF THE RESIDENTS IN THE IMMEDIATE AREA WHO ARE POSITIVE THAT THEY HAVE SEEN THIS DOG WITH ANOTHER COUPLE WHO DO NOT FRATERNIZE WITH OTHER DOG OWNERS IN YOUR AREA. I HAVE BEEN TOLD THAT THIS DOG IS NEW TO YOUR AREA AND I ASK THAT YOU KEEP A CAREFUL WATCH FOR HER, PLEASE.

DESCRIPTION:

LADY IS 3 YEARS OLD AND WEIGHS APPROX. 50 TO 60 LBS. SHE IS OF THE SHEPPARD COLLIE VARIETY WITH A BLOND COAT AND A WHITE UNDERBELLY. SHE ALSO HAS WHITE MARKINGS ON ALL OF HER FEET. LADY IS A VERY LOVING ANIMAL WHO LOVES TO BE AROUND PEOPLE AND THRIVES ON CONSTANT ATTENTION. SHE DOES NOT BITE, BARK OR GROWL AND IS VERY WELL TRAINED AND LISTENS VERY WELL.

寻找爱犬

4月15日，本来坐在我的灰色宝马车后座上的小狗不见了。肖像如图，其大腿右侧标牌号码为303。如果您看见它或者有它的信息，

请给我们打电话

464839（办公室）464822（住处）

定重谢！我们非常想念它！

（这只小狗据说是在主人宝马车后座上被偷走的！）

PERDU

TATOUÉ HMM 303 CUISSE DROITE

Le 15 avril 93 notre voiture BMW grise a été volée à LOURDES avec à l'intérieur notre petit chien Yorkshire. Il s'est peut-être échappé... Si vous l'avez **VU** ou **RECUEILLI**

TELEPHONEZ-NOUS

- Bureau **46.48.39**
- Domicile **46.48.22**

Il nous manque terriblement

TRES GROSSE
RECOMPENSE

LE MOINDRE PETIT RENSEIGNEMENT PEUT NOUS ETRE UTILE

MERCI

© IMP. MICHOT - JONZAC - 46.48.24.67 - RM 396 77 17 - NE PAS JETER SUR LA VOIE PUBLIQUE

This Yorkshire puppy was sitting in the back seat of the owner's stolen BMW.

法国

重金悬赏！
寻找爱犬“格雷西”

她重 85 磅，是只黑褐色杜宾小母犬！
最后一次见到她是在 1997 年 7 月 18 日 52 号大街和肯博克大道附近。
“格雷西”曾阻止过入室抢劫，但在追击窃贼进入街区后走丢了。
我们非常想念她，如果您见到了她，请尽快和我们联系！

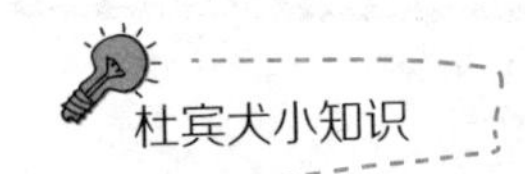

杜宾犬即笃宾犬，是一种短毛犬。原产于德国，它是根据培育这一品系人的名字路易斯·杜宾曼先生命名的，是所有品系中最富智慧的一种。身高一般在 68~72 厘米之间，体重一般在 40~45 公斤之间，寿命一般是 10~14 年。杜宾犬是军、警两用的犬种。

LOST DOG:
"GRACIE"

85 lbs. Female, Black And Tan Doberman

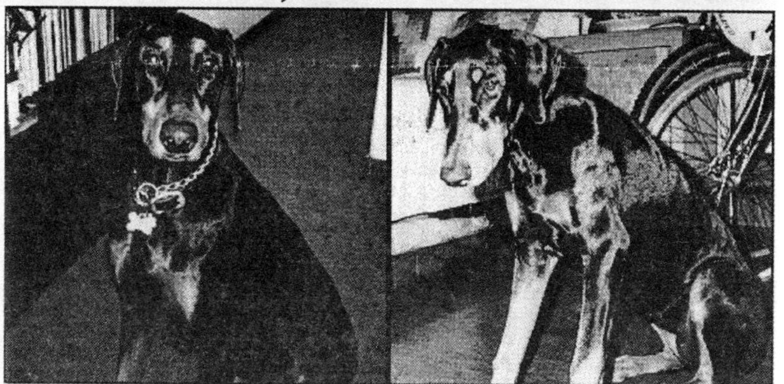

HUGE REWARD!

Last Seen The Morning Of 7-18-97

At 52ND AND KIMBARK

Gracie Foiled An Attempted Burglary, But Then Chased The Would-Be-Robber Into The Streets

寻找可爱小狗

你看见过这只可爱的小狗吗？

他是一位美国公民。他的家在佛罗里达。
如果您发现了这只可爱的小狗，请联系琳恩小姐。

她会给你丰厚报酬的。

Lost Dog

HAVE YOU SEEN THIS DOG?

He is a U.S. Citizen and needs To Return To FLORIDA.

Please Call
LEAVE MESSAGE

$REWARD IF RETURNED$
(LOCALLY)

加拿大安大略省

我最爱的小狗“托托”丢了 请好心人帮帮她

她是一只卷毛型可卡犬，有着金红相间的皮毛。她今年已经 12 岁了，高约 46 厘米，没有尾巴。她是在 1995 年 2 月 28 日离开我的，走的时候还戴着一个红色项圈，狂犬病免疫牌为 #1047。而且托托是只双目失明的小狗，我的地址是恒水街 168 号。
（托托在一场暴风雨中失踪，最终被找到，安然无恙。）

可卡犬小知识

可卡犬分英卡和美卡，原产地分别是英国和美国，又称猎鹬犬。猎鹬犬（spancial）是“西班牙之犬”的意思，从古法文“espaignol”引申而来。该犬性情开朗，聪明理性，工作认真负责，可是有时会表现得非常顽固，容易激动和兴奋，其尾巴一直剧烈地摇摆，在行动和狩猎时尤其明显。四肢和耳上的毛必须经常修剪和梳理。为了避免肥胖，必须保持足够的运动量。如给予足够的爱护，此犬很容易成为忠实的伴侣，在世界各地极受欢迎。

LOST DOG

COCKER SPANIEL TYPE

MISSING 2/28/95

PLEASE HELP - TOTO IS BLIND

She's about 18" tall, no tail, blond/red wavy hair, 12 years old.
Red cloth collar, Rabies Tag #1047.
Address: 168 Walter Street

Toto was lost during a snowstorm. She was eventually found, alive and well.

重谢 1000 美金
急寻爱犬“班吉”

爱犬“班吉”是一只杏色梗类犬，重约 8 公斤，身长 60 厘米，身高 30 厘米。我的班吉就像一部迪斯尼电影《丛林赤子心》中的主人公班吉那样勇敢且富有爱心。我非常急切地想要找到它。若有它的消息，请给我致电，重谢哦！

RECOMPENSA

DE $1000.00

POR ENCUENTRO DEL PERRO BENJI

Benji tiene pelo greñudo de color beige. Mas o menos pesa ocho kilos y mide 60 cm de largo por 30 cm de alto. Benji es de tipo Terrier Tibetano y parece igualito al Benji de la pelicula Disney de su nombre. Si tiene información sobre Benji, llame al telefono

墨西哥

“细奇”被偷了

——请不要随意撕掉或无视这张海报

“细奇”是一只纯正的雌性法兰克福腊肠犬。她有着可爱的红色短毛。喊她的名字她会回应你。她很小巧可人，只有 8 磅重，刚 1 岁半。

“细奇”在 1996 年 4 月 6 日星期六的早上，在纽约市百老汇第 105 大道上被两个 19 岁左右的西班牙男子盗走，很可能已经被其用登报或广告的形式出售，也可能被遗弃在城市的某个角落。

若您有任何关于我的灵魂伴侣“细奇”的消息，请立刻通知我，我将不胜感激并以 600 美元现金作为酬谢！

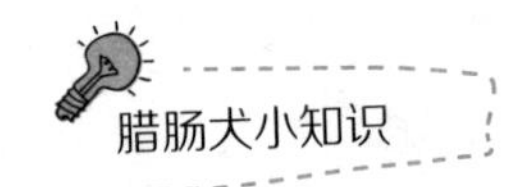

腊肠犬小知识

腊肠犬（Dachshund）是一种短腿、长身的腊犬，其名字源于德国，原意“獾狗”。此品种被发展为嗅猎、追踪及捕杀獾类及其他穴居的动物。腊肠犬属活泼、勇敢的狩猎犬种，是唯一会抓老鼠的犬种。腊肠犬可以分为标准型和迷你型两种。

"Stretch" Was Stolen, April 6, 1996

PLEASE Don't Remove this poster or dis it.

Female
Daschound
"Weiner Dog"
Pedigreed

Very small
Looks like "pup"

Tiny,
Answerts to
Name "Stretch"

Red Hair -
Short Dk.

Still Missing
April 21, 1996

Cutest dog in
NYC
8 lbs

Small Feet

Stolen Dog ("Stretch")

$600 Cash Reward

Stolen Daschound (weiner dog) named "Stretch". Red color, short hair, female, unspayed, 1 1/2 yrs. old. Miniature size, 8 lbs. Pedigree. Taken by two hispanic males, approx 19 years old.

Dog stolen early Saturday morning April 6, 1996

from 105th and Broadway, possibly sold on street, Through newspaper ad, or abandoned, at any city location. STILL MISSING!

NO QUESTIONS ASKED

Please Help if you have information about Stretch, my soulmate.

美国纽约

¡¡¡寻找走丢的小狗!!!

我的白色小狗（如图）走失了。
他是在1998年4月29日走丢的。如果您见到了他，
请拨打电话4593311
或
2073625

→ P56

¡¡¡PERDIDO!!!

PERRO PEQUINÉS ALBINO
RESPONDE AL NOMBRE DE YACKY.
29/4/98 TARDE

TELF. DE CONTACTO: **459 33 11 (MAÑANAS)**
207 36 25 (TARDES)

西班牙

被劫车时宠物狗一同被抢

1994年6月13日

美国旧金山市的警察正在搜捕一男一女两名劫匪，他们打劫了一辆庞蒂亚克小轿车。当时他们用枪指着车主，强迫其下车，然后驾车逃走，车上10岁大的宠物犬“克里奥”一并被带走。
车主芭芭拉·戴维斯告诉警察这起劫车事件发生于周一下午，18号街和波特雷罗大道交叉口。他回忆说当时他正在停车，有个男人从后车门进入车内，用枪指着他的头让他下车。当车主从他的黑色雪佛兰爬出后，一个女人钻进车后座，然后两人驾车离去，当时“克里奥”正坐在车后座上。

这辆车是加利福尼亚的牌照，号码为2MXW233。“克里奥”是一只沙皮犬，有着浅褐色皮毛，后腿处绑有一个深棕色的皮圈。跟别的动物能友好相处，见到人有点害羞。

P58

沙皮犬小知识

沙皮犬，英文名Shar Pei，产于中国广东南海大沥镇一带，是世界名种斗狗之一。它一般身高46~56厘米，体重为22~27公斤，外形为四方形。其体形独特：头似河马，嘴似瓦筒，三角眼，舌苔青蓝，幼年时全身皮肤充满褶皱，故称之为沙皮狗。标准外形：背短，身体娇小，头部比身躯大，黑鼻和深蓝色舌头，面部及全身的皱纹越深越好，耳小而下垂。毛短而刚硬，外表似乎神情忧郁，充满哀怨、凝重的沙皮犬，其实心情非常开朗、活泼、顽皮好动。

LOST DOG DUE TO CARJACKING IN S.F. 6/13/94

Man, woman sought in carjacking

SAN FRANCISCO Police are looking for a man and a woman who stole a Pontiac Transam from its owner at gunpoint, making off with the car and the owner's 10-year-old dog, Cleo.

The carjacking occurred Monday afternoon at 16th Street and Potrero Avenue, Sgt. Barbara Davis said. The owner told police he had been in his parked car when a man opened the unlocked passenger door, slid in and told him at gunpoint to get out.

As the owner climbed out of the 1982 black Transam, a woman climbed in the passenger side, and the couple sped off, with Cleo the Sharpei sitting in the back seat.

The car has a California license plate of 2MXW233. Cleo is fawn-colored, with a dark brown circle on her left hind quarter.

Compiled from Examiner staff and wire reports

NAME: CLEO
BREED: SHAR PEI
AGE: 10 YRS OLD
DESCRIPTION: Brown with small dark-brown circle on left side.
Animal friendly/people shy

美国加利福尼亚

狼狗走丢了
请致电和我联系
313-8211
重金酬谢！

LOST
WOLF
CALL
313-8211
REWARD

美国爱达荷州

急切搜寻——我家犬犬苏珊

重谢 10000 美元

提供线索者奖励 250 美元

（黑色雌狮子狗，重约 6 磅，年方 2 岁）

苏珊于 12 月 23 日在梅尔罗斯和罗瑞尔大街交叉口附近走丢。

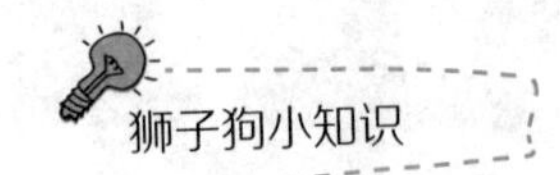

狮子狗小知识

狮子狗是一种古老的宠物狗，来自中国，是很受人喜爱的贵族宠物。狮子狗体重为3~6公斤，站立时体高15~23厘米。它天性温顺活泼，喜欢家庭环境和整洁的室外环境，在灯下它的眼睛闪闪发亮。中国人也叫它们为福狗，许多艺术作品里描绘过它们，受到人们的广泛推崇。由于它们长得像狮子，所以被认为是守护神。

DESPERATELY SEEKING SUSAN

$10,000 REWARD

OR $250.00
FOR INFO LEADING
TO HER RETURN.

BLACK FEMALE POODLE

(aprox. 6 lbs. • 2 yrs. old)

MISSING SINCE DEC. 23

FROM MELROSE & LAUREL AVE. AREA

美国加利福尼亚

奖励 1000 美元

（绝对不是开玩笑）

致收留我家“泰迪”的您：

我知道他很可爱，一定对您也具有极大的诱惑力。但对我来说，绝不仅仅是这样。他对我来说就像孩子一样重要，同时还是陪伴我将近七年的最好的朋友。他是一只 6 岁的雄性卷毛比熊犬，有着棕色的眼睛和鼻子，俏皮的白色毛发，您也觉得很可爱，对吧？

可惜的是他在 2 月 7 号那天走丢了，您一定无法想象在没有他的日子里，我生活得多么痛苦和无助，每次当我打开门发现房子里没有他的时候，我感觉我的心都碎了，总是情不自禁潸然泪下。

P64

请把他还给我！

我的电话是 549-3549。

$1,000

REWARD

NO QUESTIONS ASKED

To the person who has Teddy: I know he is irresistibly adorable, but he is more than that to me--he is like a son. My life has been absolute torment since he disappeared. Every time I walk through my front door into an empty house my heart breaks a little more and tears stream down my face. He has been my best friend for nearly 7 years...

"TEDDY"

LOST 2/7

6 YR. OLD NEUTERED

MALE BICHON FRISÉ

WHITE CURLY HAIR

BROWN EYES & NOSE

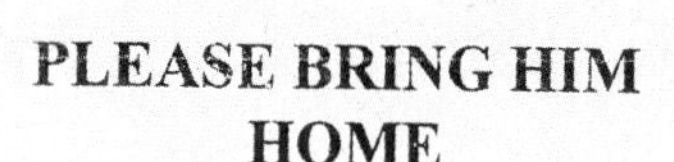

PLEASE BRING HIM HOME

IF YOU HAVE TEDDY OR SUSPECT YOU KNOW WHO DOES
PLEASE CALL ME

549-3549

美国加利福尼亚

急寻！急寻！

他叫"麦克"。

如若您见到他并将他还给我，

酬金：2000克朗

我的电话是371504。

→ P66

Pozor. Pozor

Stratil se pes
srnčí ratlík
rezavé barvy
slišící na jméno

Miki

ODMĚNA: **2000** Kč

telefon: 37 15 04

Cats

猫 哺乳动物，面部略圆，躯干长，耳壳短小，眼大，瞳孔随光线强弱而缩小放大，四肢较短，掌部有肉质的垫，行动敏捷，善跳跃，能捕鼠，毛柔软，有黑、白、黄、灰褐等色。种类很多。

送回爱猫者

奖赏50美元

我的猫猫不见了

“布丁” 雌性虎斑猫（14磅重）

10岁 黑色前爪

于1996年1月13日周六在第三大街和自由大街交叉口附近走丢。如果有人见到“布丁”，请联系海德·乔先生或者直接把它送到2号街3公寓339室。

没有戴项圈，因为一般她都不出门的。

WANTED: #50 reward

#50- reward

if found.

LOST CAT

Female tabby, rotund (14lbs)

"Pudding" 10 yr old

front/back claws

Lost on Saturday the 13th of Jan. 1996.

around the 3rd st/Liberty intersection

if found, call John Heider

339 3rd St A2.

No collar, as she's normally inside.

美国密歇根州

寻猫启事 重金酬谢！

“米奇”拥有白蓝相间的眼睛，
头上有黑斑。

这是我的爱猫“米奇”。
如果见到它的话请和蒂娜联系。
非常感谢！

→ P72

LOST CAT

THIS IS MY SWEET CAT MIKE.

PLEASE TAKE THIS HOME + PUT IT UP! Thanks, TINA.

REWARD!

White•Blue Eyes

Black smudges on head

美国密歇根州

“埃尔维斯”不见了！！！

我的爱猫是一只英俊的雄性混种暹罗猫，非常听话。他现在需要治疗，请一定留意！最后一次见到他是在7月11号，在克里斯蒂娜大街上。如果您有他的消息，请拨打电话6224911找苏珊妮，或拨打622299。万分感激！

→ P74

ELVIS
IS MISSING!!

He's a male CROSS Siamese, very friendly cat. He needs medical attention now! REWARD!! Last seen July 11th on E. CHRISTINA St. Call Suzanne 622-4911 OR 622-2999

THANK-YOU

加拿大安大略省

我的小猫咪走丢了！

- 土生土长的本地猫
- 今年满1岁啦
- 它有着黄色毛发，尾巴的毛尤其浓密。
- 您叫它的名字“卡玛”，它会回应您。

如果您有缘与它相遇，请与我联系。

TEL: 7198417

→ P76

Missing

- local breed
- 1 year old
- ginger cat with bushy tail
- answers to the name of "Karma"

If found, please contact:
Ifzan, tel. no:- 7198417

马来西亚

寻猫猫

我的爱猫走丢了。它是波斯猫和虎斑猫的混血猫，尾巴蓬松，没有戴项圈。如果你找到它，请把它送到米洛格的B单元。谢谢。

→ P78

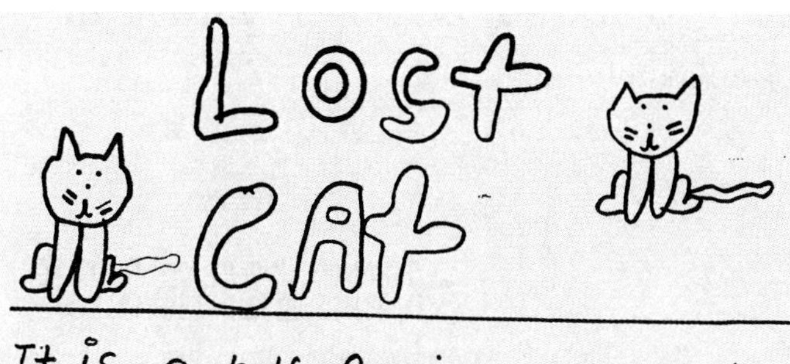

It is a half Persian cat and half tabby. It has a fluffy tail. It has no coller. If you find this cat Please return it to 10 minogue units. Thank you. Please find my home.

社区新闻

寻找可爱的小猫凯蒂

5 个月大的小猫流浪中……

一只仅有 5 个月大的可爱的虎斑猫离开了主人鲍勃·雅各比的家（西雅图枫叶街区 98 大街 801 室），在周四大胆地出走了。这只小猫现在仍在流浪，给这个原本宁静的社区带来了一些乐趣（或灾难）。

在 11 月 11 日，周四的某时，他开始了大胆的出逃。这只 5 个月大的雄性虎斑猫有着黑棕相间的毛发，一直被认为是个“讨厌的家伙”。他可能戴着一个项圈（除非他自己把它卸了下来）。

根据主人的介绍，这只小猫尤其喜欢在浴盆和马桶处玩耍，他喜欢看放水时水流在他眼前形成的旋涡。当这只淘气的小猫崽在附近的时候，气球、绳头和袜子都可能会遭殃。最后一次看到他是在罗斯福大街附近。

这个讨厌的家伙后来在罗斯福大街和 98 号大街附近被人看到，当时他正在兴致勃勃地看人洗车呢！

这个家伙在罗斯福大街和 97 号大街莱根公寓处找到藏身之所

在那里他找到一条进出公寓大楼的捷径。在那里待了几天后，他的流浪瘾发作，在 12 月 16 日星期二又消失了。

他的主人，鲍勃·雅各比，刚刚和那所公寓楼的住户取得了联系，却只得到他已经离开的消息。“等我找到那个小无赖，一定给他戴上身份铭牌。”鲍勃·雅各比声称。天就要黑了，可鲍勃·雅各比还得赶往下个地点去寻找这只小猫。

P80

如果您看到凯蒂，请联络鲍勃·雅各比！

EXTRA! Neighborhood News EXTRA!

© 1993 by Friends of Twerp; Phone: 525-0608

MISSING KITTEN!

WANDERLUST HITS 5 MONTH OLD KITTY

In a daring midweek escape, a clever 5 month old tabby kitten has left the home of his owner, Bob Jacoby, at 801 N.E. 98th Street in the Maple Leaf neighborhood of Seattle. The kitty is now at large, and is charming (or terrorizing) this formerly peaceful neighnorhood.

The daring getaway occurred some time around Thursday, November 11th. The kitty, known as "Twerp", is a black and brown 5 month old male tabby, possibly wearing a flea collar (unless he has ditched it).

According to the owner, Twerp has a strange affinity for bath tubs and toilet bowls, watching the water swirl around and disappear down the drain before his eyes. Balloons, pieces of string, and socks are all in jeopardy when this cute little terror-meister is nearby!

A recent photo of "Twerp", obviously studying one of his potential escape routes.

LAST SEEN NEAR 97TH AND ROOSEVELT WAY

Twerp was later sighted near 98th and Roosevelt Way, where he watched and lent moral support for several hours to a man working on his car.

Twerp later sought refuge at the Legend House Apartments at 97th and Roosevelt Way, where he had charmed his way into one of the apartments. He was content to stay there for several days until wanderlust hit again, and he disappeared some time during Tuesday, November 16th.

His owner, Bob Jacoby, had just finally been able to make contact with the woman in the apartment, only to find that the clever Twerp had slipped out that afternoon! "That little rascal is going to have an I.D. collar on him as soon as I find him!" exclaimed Mr. Jacoby as he headed out into the night on another search party.

If you see this kitty, please call Bob Jacoby

美国华盛顿

爱猫“黑人”不见了

150.00

作为酬谢！

黑色雄性小猫，戴黄色项圈，
耳朵上有斑纹，尾巴下面有伤疤。
请将他送回家，无论是生是死。
请和我联系！

Tel：385-4098

WANTED

DEAD OR ALIVE

$ 150.00 $

REWARD

'BLACKIE'

BLACK MALE, YELLOW COLLA

TATOO ON EAR

SCAR ON TAIL BASE

PH. 385-4098

寻找我的猫咪

大名“猪猪”，

是只公猫。

外皮橙色加白色，

耳朵上有两个疙瘩。

→ P84

LOST CAT
H's NAMe is
Piggy. He is
ORANge AND
WHITe AND HAS 2
BUMPS on his Ear.

加拿大新斯科舍省

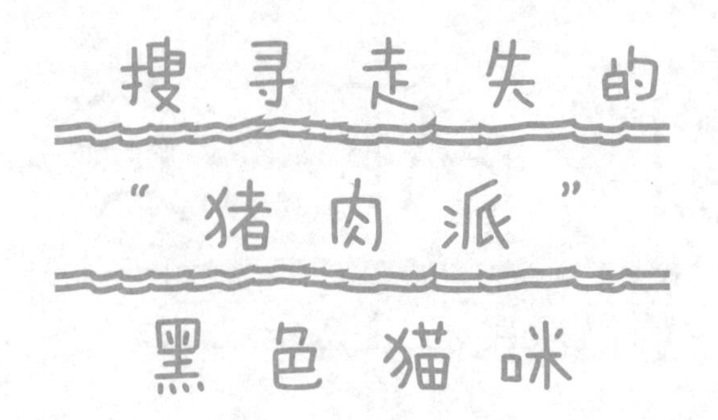

我们非常爱他。
非常想他。

请与英国伦敦艾菲尔德大街的罗伊斯顿·杜·莫里艾·莱贝克联系！

P86

LOST

"PORKY PIE"

BLACK CAT

LOVED

LOVED

ROYSTON DU MAURIER-LEBEK.
IFIELD RD. LONDON SW10 9AD U.K.

英格兰

我的小猫咪到底跑到哪里去了？你们见过吗？

她是只橙色的母猫，毛皮光滑，性情暴躁。

请帮忙寻找，谢谢！

（注：这只“猫咪”有四对乳房，性格就像个小老虎，很少叫，喜欢抓小孩子。）

J'entends du bruit... je me lève... plus de petite Nonette! Où est donc passée cette créature de rêve? Des yeux verts, 4 paires de seins et des poils si doux!!! Recherche donc chatte européenne blanche et orange. Tempérament de tigresse... ne ronronne que très rarement et aimant griffer les enfants! Appeler le

.31.11.79
En France!

This "pussy" has "four pairs of breasts...the temperament of a tiger...rarely purrs, and likes to scratch children".

法国

你们见过我的小猫吗？

在星期日，我可爱的宝贝走丢了。
她是一只可爱的白雪公主。
她听不到声音，很可怜。
请好心人帮忙找到她，并与我联系。
非常感谢！

→ P90

Have you scene my CAT?

I lost my CAT sunday..
She is white and is also deaf. Someone please
c · her and coll

CAT CAT CAT CAT

加拿大魁北克省

寻猫启事

名字：Ringer

特征：虎斑纹
卷尾巴
很热情

我真的很想他，我还想和他一起过圣诞节呢！

谢谢！

→ P92

注：“ringer”在英文中是“套环”的意思，故猜测主人给小猫起此名的原因是猫咪身上的花纹似一圈一圈的圆环。

Lost Cat

Name: Ringer

Description: tabby color
Ring tail
affectionate

I Really miss him
and I would Like him home
for Xmas.

thank You

寻找猫咪

急寻雄性猫咪一只，身穿橘黄色大衣，约3~4岁。

请速与我联系：274-736。

喵喵！

→ P94

Lost Cat

orange cat lost

Male 3-4 years Old

Please phone: 274-736

Ceaser

加拿大魁北克省

我走丢了！！

记忆中的最后一个地点在博伊德·克里夫顿的加富尔大街附近（1995年2月23日）。

我的名字叫萨米，在北岭市地震中幸免于难，现在需要您的帮助，让我继续活下去。请检查一下您的地下室或车库吧。

我很友好，1岁半了，有着黑白相间的短毛，看起来像是燕尾服。主人非常爱我。

一定重谢！

I'm lost!!

Last seen 2-23-95
near Boyd,
Clifton, Cavour Ave

My name is
Sammy. I
survived the Northridge earthquake
Please help me survive this.

Check your basement or
garage. I'm friendly.

I'm black + white, shorthair,
1 1/2 yrs. old, "tuxedo" markings.
My owner loves me.

REWARD!

美国加利福尼亚

寻找我最宝贝的猫，9 月 13 日至今未归

你见过我的“托尼”吗？他可是我最喜爱的宠物猫。他是只大块头的虎斑猫，10 岁左右。拥有橙色和卡其色相间的毛色，背部有白色斑点，白色的爪子。毛发不长也不短，颈部有蓝色项圈和英国皇家兽医所发的健康标牌。

丢失那天，我们出门在外，于是把他放在了朋友家，他很可能试图自己回家。朋友家的位置在杜邦和司帕蒂娜街区。他是只非常害羞和胆小的猫，我们非常想念他，请帮我们找找他。

注意：9 月 20 日他曾被人看到在豪兰路上游荡，身上有伤。请检查您的车库和任何他有可能藏身的地方，谢谢！如果您告诉我他的位置或带他平安回家，我们将会支付您 100 美元的报酬以示谢意。

P98

MISSING SINCE SEPTEMBER 13

HAVE YOU SEEN TAWNY, OUR MUCH LOVED MISSING CAT?

Note: On September 20 he was spotted on Howland Avenue, possibly with an injured leg.

Please check garages, and anywhere he could be hiding. Thank you!

DESCRIPTION:

LARGE TABBY - APPROXIMATELY 10 YEARS OLD; ORANGEY/BEIGE WITH WHITE SPOT ON BACK; WHITE PAWS; MEDIUM HAIR; BLUE COLLAR WITH 2 KEW BEACH VET TAGS

He was staying at a friends home, (Dupont/Spadina Rd.) while we were away and may be trying to get home (the beaches). He is shy and is probably very scared. We miss him very much - please help.

WE ARE OFFERING A $100.00 REWARD FOR INFORMATION LEADING TO HIS SAFE RETURN.

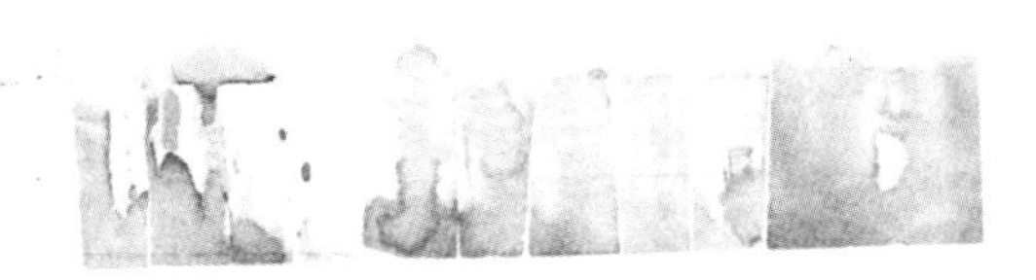

加拿大安大略省

“克利奥”惨遭诱拐！！

“克利奥”是只蜜色的成熟小母猫，毛发光滑明亮且漂亮动人。

她没有尾巴。

6 月 21 日在大学路和贝弗莉大街交界处，她被人从车中劫走。

克利奥由于健康问题必须严格控制饮食，我们非常爱她，非常想念她。

请帮我们找到她。

CLEO THE CAT ABDUCTED!!

Cleo is a honey-coloured, mature cat, very fluffy, and has

NO TAIL

She was picked up in a car from College and Beverly Streets on July 21

Cleo is on a restricted diet because of a health problem. She is loved very much and missed terribly

Please help us find Cleo

寻找爱猫

一只大龄且牙齿掉光的橘黄色波斯猫走丢了，

他有着扁平的脸庞，皮屑问题很严重。

他是一位很重要的家庭成员，名叫“诺曼”。

3月15日星期日下午两点在西夫韦超市走丢，

请好心人帮忙找到他！

我们将以200美元表达谢意。

电话：661551

P102

波斯猫小知识

波斯猫（Persian）是最常见的长毛猫。它是以阿富汗的土种长毛猫和土耳其的安哥拉长毛猫为基础，在英国经过100多年的选种繁殖，于1860年诞生的一个品种。波斯猫乃猫中贵族，性情温文尔雅，聪明敏捷，善解人意，少动好静，叫声尖细柔美，爱撒娇，举止风度翩翩，天生一副娇生惯养之态，给人一种华丽高贵的感觉，故有“猫中王子”、“王妃”之称。并且体格健壮有力，躯体线条简洁流畅：圆脸、扁鼻、腿粗短，耳小、眼大、尾短圆。波斯猫的背毛长而密，质地如棉，轻如丝；毛色艳丽，光彩华贵，变化多样。

LOST CAT

AN OLD AND TOOTHLESS, ORANGE
PERSIAN WITH A DANDRIFF PROBLE
AND A FLAT FACE WHO IS AN
IMPORTANT FAMILY MEMBER.
PLEASE HELP TO
FIND "NORMAN"
(MLK
(3-15)
– DISAPPEARED SUN. @2pm @SAFEWA

REWARD
$200.00

(NO QUESTIONS ASKED)
PLEASE
CALL # 661-551

美国俄勒冈州

我将用 100 美元作为酬谢！

我的猫咪走丢了

黑毛的“汤姆”，

颈上戴着一个黄色项圈。

如果你见到这只猫咪的话，请给我打电话。

我的号码是 0181-910。

→ P104

-REWARD £100-

CAT LOST

BLACK TOM
WITH
YELLOW COLLAR

LEAMINGTON ROAD VILLAS
PLEASE CALL 0181-910
IF YOU HAVE SEEN HIM

急寻爱猫

1996年4月21日，我的爱猫"普斯·普斯"走丢了。
她通体黝黑，胸口有白色斑点。
我猜她现在很有可能还在某处房顶上。
我这样推测是有原因的——她是从克吕贝克街78号的屋顶上逃走的。麻烦各位好心人抬头往高处看看，
也许"普斯·普斯"就在你上方。多谢！

→ P106

Wir Vermissen
unser Kätzchen
seit dem21. 4. 96.
Sie ist ganz Schwarz
mit einem weissen
fleck auf der Brust.
Ihr Nameist Puspus. Evtl.
Ist sie in einem estrich, da sie
übers Dach der Klybeckstr. 78
weg gegangen ist. Bitte
Schaut überall nach. Danke.

Her name is Puss Puss.

寻找爱猫

爱猫名叫"菲利克斯"，雄性。
他有8磅重，戴着棕色项圈，没有铭牌。
灰色毛发中带有浅灰色的条纹。
他不幸在地震中走丢。我非常想念我的小猫，如果您见到了他，请和我联系！

重谢！

LOST CaT

during Earthquake

NAME: FELIX (male)

Color: Grey with light grey stripes. 8 pounds.

BROWN Colar, NO TAGS

OWNER Misses Him VERY MUCH. PLEASE CALL

$ $ REWARD $ $

美国加利福尼亚

急寻爱猫

你见过"卡杰"吗？

他是只大约3磅重的猫咪。

有着黑色毛皮，前胸是白色的。

请帮忙寻找他，谢谢！

（在这张海报张贴之前，一只和"卡杰"长相相似的猫曾向邻居讨食，但是由于她在减肥，邻居并没有给她喂食，结果很不幸，一周后发现她被车撞飞后死于灌木丛中。）

→ P110

Before this poster appeared, others of the same cat asked neighbors not to feed her because of her strict diet. She was found dead a week later. She'd been hit by a car and died in the bushes.

荷兰

黑白相间的猫走丢了

这只猫的前爪多了几个指头。

请查看一下您的车库或者小屋，也许它就在那里。

如果发现了它请联系亚当。

若它给您带来不便，敬请谅解。

LOST
BLACK &
WHITE
CAT

WITH A FEW EXTRA FRONT TOES

PLEASE TAKE A LOOK
IN YOUR GARAGE OR SHED

IF YOU FIND SCOOTER PLEASE CALL
ADAM

加拿大安大略省

请帮忙寻找小猫
他的主人是
达姆 和 阿努克

小猫八个月大，
拥有红棕色的腹部，
白色的脖子和爪子。

定重谢！

→ S.V.P. BESOIN AIDE ..

TAM ET ANOUK

→ CHERCHENT

LEUR PETIT

CHAT

ADORÉ ..

MIMY.

SI CELA PEUT AIDER RÉCOMPENSE

MERCI BEAUCOUP.

DE TAILLE QUASI ADULTE ≃ 8 MOIS

ROUX TIGRÉ

VENTRE BLANC

(COU ET PATTES)

法国

急寻爱猫

这是杰瑞，也叫“大胖猫”。

非常可爱。

于3月29日星期一那天在杜邦和斯帕蒂娜街区迷路。

我非常着急。

如果您有看到它，请拨打电话，

我将以50美元来表达我的感激之情！

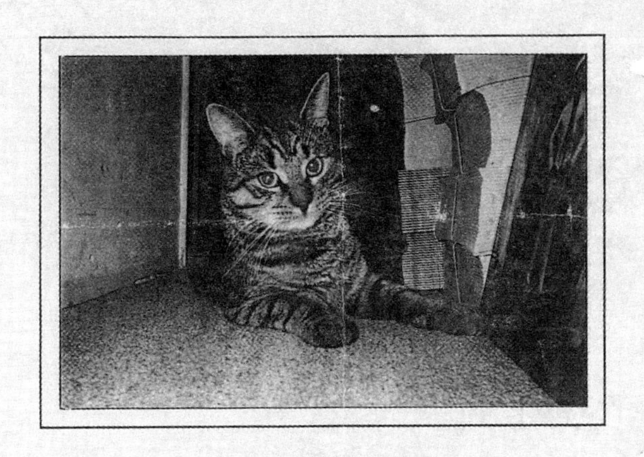

This is Jerry
also called" big fat kitty "
went astray on Mon.March 29th
Dupont / Spadina area.

If found please call

$50.00 Reward

悬赏 30 美元

搜寻爱猫

他是一只 3 岁大的雄性暹罗猫，已阉割，身上有着细长的深棕色和白色相间的花纹，和主人感情深厚。

他是在 11 月 22 号星期一，在女王大街和国会大街的交叉口走失的，走丢时戴有项圈。

我非常非常想念他，望好心人见到他跟我联系，谢谢！

SIAMESE CAT.
3 YEAR OLD NEUTERED MALE. SEAL-POINT MARKINGS (DARK BROWN & WHITE) VERY SLENDER VERY AFFECTIONATE. LAST SEEN MONDAY NOV 22. QUEEN AND PARLIAMENT STS. WAS WEARING COLLAR WITH I.D. TAG. SORELY MISSED.

寻找主人

我的名字叫“吉默”。今年5岁了！

我是一条黑白相间的小猫。

戴有一条红色项圈，性情温顺。

我自从6月26日后就再也没有回过家了。呜呜！

若你见到我，请告诉主人我的位置，他们一定很担心我。

电话：485617。

"My name is Gummo."

法国

通缉此猫

你见到过“科尼利厄斯”吗？他是只中等个头的灰色虎斑猫，爪子和前胸都是白色的。他没有尾巴，走起路来神气傲慢。

他在9月22号星期二走丢了。他正生着病，如果不按时给他医治的话，他会有生命危险的。

他在央街和卡尔街区生活多年，附近的邻居应该能认出他。

如果你有关于“科尼利厄斯”的消息，请与我联系。万分感谢！

P122

WANTED

Have you seen Cornelius? He has been missing since Tuesday, September 22nd.

Cornelius is on medication and must be treated immediately or he will die!

You will recognize Cornelius as having been a fixture at Yonge and St. Clair for many years.

Description: He is a medium sized gray tabby with white paws and white chest. He has no tail and walks with a slight swagger.

If you have any information to help Cornelius get home safely please contact

寻找“凯蒂·朗”

她是埃塔的办公室爱宠，3月15日星期三走丢。她是一只灰白相间的长毛本地猫，戴着一条蓝色猫牌。

如果你见到她或知道她的位置，请联系埃塔。

（凯蒂·朗一年后回来了，变得神采奕奕且有点发福，身后还跟着另一只猫。凯蒂走丢后，办公室又养了一只猫，于是后来凯蒂被送到一名员工亲戚的乡村农场居住，另一只猫也被送给别人收养。）

→ P124

MISSING

KITTY LANG,

Xtra's office cat, has been missing since Wednesday, March 15th. She is a gray & white long-haired domestic and is wearing a blue identification tag. If you've seen her or have any idea as to her whereabouts, please contact Xtra

kitty Lang came back a year later, well groomed and overweight, with a second cat. Since a new office cat had taken residence, kitty moved out to the country farm of a relative of one employee and the second cat was adopted by another.

若将他平安送回，
主人将以 100 美元表示感谢。

寻猫启事

“金吉”是一只橘黄色虎斑猫，体重 10 磅，非常友好温顺，家住布伦瑞克－阿尔斯特地区。如有他的消息，请联系我们。

（注：“金吉”已经找到，谢谢各位的帮忙。）

→ P126

CAT

GINGER HAS BEEN FOUND!
Thanks For Your Help.

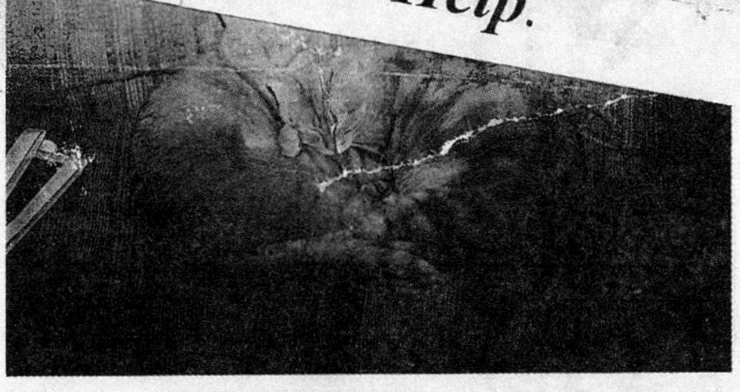

GINGER (left)

Ginger is an orange tabby cat who weighs ten pounds. He is very friendly and greatly mi d. Ginger lives in the Brunswick – Ulster area. If yo ave any information please phone $100. reward for his safe return.

$100 REWARD

猫咪，你在哪里？

她是一只纤小、甜美、黑色的小母猫，名叫“格瑞泽贝拉”。相信她现在正在墨尔本北部的某处可怜地游荡。如有发现她的踪影，请联系尼基·坎贝尔。

多谢了！

请帮帮我！喵！

（注：“格瑞泽贝拉”这个名字来自于百老汇音乐剧《猫》中“魅力猫”的意思。）

→ P128

MISSING

Small, sweet, black, female cat
Answers to the name of
' Grizzabella '
Believed to be somewhere in the vicinity of
North Melbourne
If found, please contact Nicky Campbell

THANK-YOU

Grizzabella was the name of "The Glamour Cat" in the Broadway musical "Cats".

寻找喜欢吮着水龙头喝水的爱猫

他有白色的爪子，一只后脚脚垫是粉色的。

他是一只成年的山猫，戴着白色围兜，跟他的白肚皮相映成趣。

他的脖子上套着项圈，于 7 月 24 日星期一在普特南附近的剑桥区，弗兰克林大道走失。

他不是一只流浪猫。我很挂念他。

若你见到他，请给弗兰德打个电话（975-0975）。

哦，对了，他的名字叫“普希金”。

P130

He likes to drink from faucets, has

white feet, with one pink paw pad (on a rear foot) and is an adult gray tiger cat with a white bib and belly. He was wearing a flea collar when he wandered off **Mon., July 24,** from Franklin St., Cambridge, near Putnam.

He is not a stray, and we miss him. If you see him, please give Fred a call at 975-0975. His name is Pushkin.

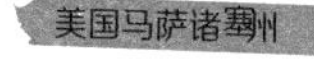

寻找黑白短毛猫

我的一只黑白短毛猫走丢了。

他的遗失地点在片打东街503号。

如有发现请致电，定重谢哦！

→ P132

寻找

黑白短毛貓

遺失地點：

片打東街503號.

如有發現請致電

恩有報酬. $$$

加拿大不列颠哥伦比亚省

非常想念黑色大猫

他个头很大，神情慵懒，但又极其友好，

他就是我走丢的爱宠“**比夫**”。

最后一次看到他时，他正在第二大道上慵懒地闲逛。他通体黑亮，腹部有白色斑点，脚掌上戴着脚套。若您有他的消息，请跟我联系，我会一直在电话旁守候的。

附：塔格、南希和姨妈都非常想念这个大家伙。

→ P134

MISSING

ONE

VERY LARGE

VERY LAID BACK

EXTREMELY FRIENDLY

BLACK CAT

"BIFF"

A/K/A "THE BIFFSTER" OR
"THE SEAL PUP"

Last seen wandering vaguely around Second Street wearing
a GLEAMING BLACK COAT,
with a WHITE SPOT on his tummy
and THUMBS (MITTENS) on his paws

Operators are standing by to take your call

PS. Tug, Nessy & the Aunts (Unlimited) really miss the big guy

你曾经见过我吗？

$10.00 "洛佩斯" $10.00

我于4月11号星期日不幸走失。我通体黑色，脖子上戴有黑色心形项圈，尾巴轻微骨折。我的家在北部的223大街上。主人非常爱我的！

→ P136

HAVE YOU SEEN ME?

LOPEZ

$10.00 REWARD

$10.0 REWARD

I HAVE BEEN **MISSING** FROM MY HOME OF **223 AVE. E NORTH** SINCE SUN. ARRIL 10th. I AM COMPLETLY **BLACK** WITH A WHITE COLLAR WITH BLACK HEARTS ON IT. I ALSO HAVE A SMALL FRACTURE IN MY TAIL. MY OWNERS LOVE ME **VERY** MUCH.

加拿大不列颠哥伦比亚省

我的爱猫走丢了！

他是一只小公猫。请帮忙找找他，
我将以1克朗以示谢意。
我的联系电话是

42·35180

我家住在汉斯大街。

→ P138

KAT.
LL. STRIBET HAN KAT
TILLØBET 1. OKTOB.
KAN IKKE FINDE HJEM
42.35180
(SCT. HANS GADE)

丹麦

失猫招领

本人于 1994 年 11 月 26 日捡到一只暹罗猫。为了确保它回到真正的主人那里以便获得幸福安宁的生活，请失主提供至少 5 条关于这只猫的身份特征，最好包括确切的电话号码和工作地址。而且，在您领走之前务必详细描述猫的特征，还要给我一张猫的照片，日后我会按照这只猫的特征进行核对。

希望失主尽早与我联系。

→ P140

CAT FOUND

Siamese

11-26-94

In order to safeguard the welfare of the cat and ensure that it is returned to it's rightful owner, the person claiming ownership of the cat will be required to provide: five (5) forms of identification and references including but not limited to, verifiable telephone number and place of occupation. Also, a very specific physical description and photo must be given before return. I will personally check up on the cat at a later date.

你见过这只猫吗？

右边的这只已找到，左边那只仍在流浪！

如果你见到了一只猫，请核对海报图片。

我外出了一个月，回家后发现保姆把我的猫咪弄丢了。

它是我生活中极为重要的伙伴。

如果您有它的消息，请给我电话，非常感谢。

→ P142

STILL MISSING.

TAKE PHOTO IF NECESSARY.

THIS ONE FOUND OK!

HAVE YOU SEEN THIS CAT?

I was away for a month and returned to find out the sitter lost them. Please phone if you have, or have seen, my much loved pal. Thanks!

加拿大安大略省

我的可爱布偶猫于1111年12月24日走丢了。她的下巴上有白色条纹（很像长毛的暹罗猫）。9岁大，喜欢在水池旁玩耍。那一天家里发生火灾，她受到惊吓后离家出走，从此一去不回。家人都很爱她，希望有人看到她后告诉我，定重谢！

注：请检查您的车库或花园屋棚，她很有可能就在那里！谢谢！

→ P144

布偶猫小知识

布偶猫原产于美国，又称"布拉多尔猫"，是猫中体形和体重较大的一种猫。其祖先为白色长毛猫与伯曼猫，于1960年开始繁育，1965年在美国获得认可。布偶猫头呈V形，眼大而圆，背毛丰厚，四肢粗大，尾长，身体柔软，多为双色或三色猫。布偶猫全身特别松弛柔软，像软绵绵的布偶一样，性格温顺而恬静，对人非常友善，忍耐性强，对疼痛的忍受性相当强，常被误认为缺乏疼痛感。非常能容忍孩子的玩弄，所以得名布偶猫，是非常理想的家庭宠物。

Missing since Dec. 24/97. Himalyan Rag-Doll cat. Very friendly. (Like long-haired Siamese). Near peanut pond. House-fire may have frightened her. Much loved family pet. 9 years old. Reward. #4

May have inadvertently got locked in garage or garden shed.

Thank You.

White stripe under chin.

加拿大不列颠哥伦比亚省

寻找丢失的小猫

这只猫非常可爱，如果你看到它一定会喜欢上它的。

它有着灰黑相间的条纹，而它的爪子是白色的哦！

不过，4 月份的时候它在汉密尔顿北部第 300 大街走丢了。

如果有谁看到了它，请记得联系我，拜托了！

它体形有点小，像只年幼的小猫。

找到有赏哦！

→ P146

LOST CAT

LOST SAT. APR. 9 FROM
300 BLOCK N. HAMILTON.
GREY WITH BLACK STRIPES
AND WHITE PAWS. KINDA SMALL
JUST A KITTY. PLEASE CALL
IF SEEN OR FOUND.
REWARD!!

美国威斯康星州

小猫丢了！

它是一只黑白相间的小猫，

八周大，左面的耳朵弯弯的，很显眼。

我的电话是 2443463

→ P148

LOST

BLACK AND WHITE KITTEN

8 WEEKS OLD

DISTINGUISHING BENT LEFT EAR

PHONE 244 3463

失猫招领

本人于 15 号星期六那天捡到一只猫咪，如图。

请猫咪的主人打电话 746407 和我联系。

P150

CAT FOUND

SAT 15th

CALL: 746 407

英格兰

寻找爱猫

11月1日，本人的爱猫不幸丢失，我和家人非常担心。

两个爪子都是六趾
通体纯白

如果你见到他，请致电673-259。

→ P152

WITHOUT REWRITING NAMES
TEAR OFF AT PERFORATION

LOST since November 1

Our cat

Description

6 toe

All white

If you have seen he

Please call at 673-25

美国加利福尼亚

寻猫启事

我的雄性虎斑猫“凯蒂”不见了，
他才4个月大。我很挂念他！

电话：256-5631

→ P154

LOST!

4 month old
male tabby

KITTEN

please call 256-5631

寻猫启事

大名BENNR，黑白猫一只。

电话：234415

重重有赏！

→ P156

SMARRITO
GATTO BIANCO-NERO
DI NOME BENNY.
SE QUALCUNO HA NOTIZIE
PER FAVORE
CI TELEFONI 234415
RICOMPENSA!

意大利

寻找小猫约翰尼

周四晚，我家猫咪趁我们都不在家时出去偷腥。他是一只还不满1岁的公猫。奶油色的毛发上有棕灰色的斑点。蓝色的眼睛，粉色橡皮般的鼻子。他很友好，刚刚被收养了4天。我已尽全力去寻找他，可还是没有任何线索。如果您有他的消息，请联系我，我将万分感谢。

联系人：布莱德·萨拉

电话：782-2412

→ P158

I am a little stray tomcat - cream body w/ brown-gray points + a very long striped tail. Blue eyes, and a nose the coler of a pink pencil eraser. I am friendly, not quite a year old. I have only been adopted for 4 days. I just got all my shots, and was supposed to get my male cat operation, so I ran away Thursday night when nobody was home.

Please call Brad or Sarah

782-2412

THANK YOU

美国华盛顿

你有见过我吗?

我叫“皮寇”。

我有着白色的大肚子，没有戴项圈。

我不小心走丢了，找不到回家的路了。喵喵!

麻烦好心人见到我后把我送回主人身边，

万分感谢哦!

M'avez-vous vu ?

加拿大魁北克省

喵喵……喵？

寻找凯特

大名：凯特

地址：木德街 2227 号

非常可爱，如果您有见到它，请联系我，非常感谢哦！

黑色混血暹罗猫

meaw meaw meaw meaw?
LOSt Kitten
Name. Kitty
Address: 2227 E. Moodie st.
What Kind of cat. Half Siameses Half normat.
color black
meaws

加拿大安大略省

Birds

鸟 脊椎动物的一大类，体温恒定，卵生，嘴内无齿，全身有羽毛，胸部有龙骨突起，前肢变成翼，后肢能行走。一般的鸟都会飞，也有的两翼退化，不能飞行。燕、鹰、鸡、鸭、鸵鸟等都属于鸟类。

走失！
走失！
走失！

我的宠物是一只棕白相间的鸭子。

它有一个特别的名字叫“内泽·诺曼”，

可是喊它的名字时它不会理你的。

如发现或有相关线索，

请与“一只鸭子和三个艺术家”联系。

真心感谢！

→ P166

LOST

LOST

LOST

one
brown and white 'mottled'
street duck

Does not answer
to the name of
Neither Norman

if found
please call
Three Artists and a (Duck)

200 美元重金回报

我心爱的纯白色葵花鹦鹉

"史班奇"不见了。

它有着橙色面颊和黄色顶冠。

它听得懂自己的名字哦！

如有线索，请联系 662-4452

葵花鹦鹉小知识

葵花鹦鹉，英文名 cacatua galerita，原产于澳大利亚北部、东部及东南部至昆士兰岛西部，新几内亚及北部、东部岛屿等地。体长 40~50 厘米，寿命一般为 40 年左右，也有的活到 60~80 年。羽毛雪白漂亮，头顶有黄色冠羽，在受到外界干扰时，冠羽便呈扇状竖立起来，就像一朵盛开的葵花。食物以葵花籽、玉米、花生米、高粱、稻子为主，每天可加喂些苹果和少量青菜。这种鹦鹉聪明，乖巧，富有感情，叫声响亮，善学人语，深受鹦鹉爱好者喜爱。

$200 REWARD

For any information
leading to the return
of my bird to me.

LOST

All white

COCKATIEL

orange cheeks

yellow crown.

Answers to the name

"*Spanky*"

Call:

662-4252

急寻失踪南非灰鹦鹉

悬赏500美元

名字："艾希莉"

颜色：灰色羽毛和红色尾巴

身高：1尺多，双翅展开有两尺那么长

在9月10日星期一下午7点在德克萨斯州失踪了。

请拨打电话8245469和我联系。

→ P170

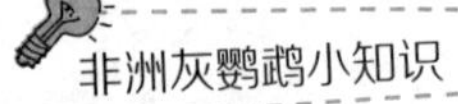

非洲灰鹦鹉小知识

非洲灰鹦鹉（学名：Psittacus erithacus）属于大型鹦鹉，是典型的攀禽，对趾型足，两趾向前两趾向后，适合抓握，鸟喙强劲有力，可以食用硬壳果。尾巴短，头部圆，面部长毛，喜攀爬，不善飞翔。非洲灰鹦鹉是已知的几种可以和人类真正交谈的动物之一，这也使得它们成为知名度最高的宠物鸟之一。主食各类种子、坚果、水果、花蜜、浆果等。分布于非洲

LOST PARROT
AFRICAN GREY
REWARD $500.00
NAME: ASHLEY
COLOR: GREY
RED TAIL
SIZE: 1FT TALL
WING SPAN 2FT
FAMILY PET
LOST MONDAY
9/18 7:00
P.M.
at Texas/19th
CALL 8245469

美国加利福尼亚

加急！ 寻鸟启事

他是一只澳洲鹦鹉，名叫贝利。雄性，黄色羽冠，白色羽毛，双颊有红色可爱斑点。

他于6月8日周六下午6点左右在雪邦街和滨海大道附近走丢。我们快要急疯了，很担心他的安全。

若有任何有关此鸟的消息，请务必联系我们。非常感谢！

澳洲鹦鹉小知识

澳洲鹦鹉是鹦鹉家族中较小的一种，其大小介于最小的虎皮鹦鹉和最大的非洲灰鹦鹉之间，与人关系好，像狗一样认主人。飞得不高，喜欢歇在人的肩膀上与主人玩耍，比较淘气，可以模仿说人话。

LOST BIRD

WANTED

ESCAPED COCKATIEL, MALE

WHITE, PALE YELLOW HEAD, RED SPOTS

LOST SATURDAY, JUNE 8, AROUND 6 P.M.

NEAR SHERBOURNE + ESPLANADE

WE ARE VERY WORRIED ABOUT HIS SAFETY
AND WOULD APPRECIATE ANY INFORMATION

雏鹰失踪了！

我的可爱雏鹰“哈里斯”在人海中走丢了。

特征：

他有着红棕色的羽毛，大黄色的脚丫子，胸脯上有灰白色的斑点。他的头部是黑棕色的，肚子是白色的。他的尾巴很个性，两腿系有铃铛和彩带。

他很勇猛，但很怕狗。

如有线索，请打电话和桑迪联系。非常感谢！

885913

→ P174

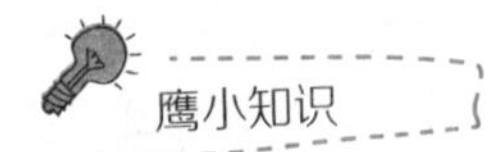

鹰小知识

鹰，鸟类，上嘴呈钩形，颈短，腿部有长毛，足趾有长而锐利的爪。是猛禽，捕食小兽及其他鸟类。种类很多，如苍鹰、雀鹰、老鹰等。

LOST

HARRIS HAWK

Juvenile (Male) Harris Hawk lost
on Ranmore Common.

Description:

Red/Brown colour. Large Yellow Feet
Mottled Brown/white on chest.
Distinctive tail,Dark Brown with White
tip and underside. Bells and Flying
Jesses' on both legs.

He is very tame, but scared of Dogs. If seen
or heard PLEASE RING Sandy.

885913

我的小鸟不见了！

她叫“布兰施”，
羽毛黄黄的，脸颊橘红的。
若你见到她，请打电话给我。

谢谢你！

P176

Missing bird!

her name is blusher

She is yellow with

ornge cheaks

if you see her call

Thank you

急切搜寻心爱鹦鹉

我们非常喜爱的红眼亚马逊鹦鹉——“飞马”，在6月3日下午大概5点50分在它寄养的“Spectrum奇异鸟寄养中心”被人盗走，我们那时在外度假。她大概7英寸长，绿色外衣，喙的上方是白色，眼睛周围则为红色，尾羽上也有点点红色。我非常担心和挂念她的安危。她跟我们共同生活了5年之久，早已是家中的一分子，没有她，我真的很难过。

如果你有此鸟的相关线索，请迅速与我们联系。928-5823。如果所拨电话不通请留言。您也可以留下匿名信息或您的姓名，我们会以此为据对您的真诚付出表达谢意。请留下充分的信息让我们得知她就是我们急切找寻的爱鸟。

亚马逊鹦鹉小知识

亚马逊鹦鹉（学名：Amazona）的羽毛大部分为绿色，眼睛虹膜桔色。黄色分布在头冠、眼喙之间和大腿处，眼睛周围偶尔也可见。头顶的黄色也叫做“帽”。在翅膀上有少许红色点缀，羽毛边缘呈黄绿色。体长34~45厘米不等。身体为绿色，带蓝的深绿色至偏黄的绿色都有，头部羽毛颜色各异，可鉴别其品种。亚马逊鹦鹉分布在中美洲、南美洲和墨西哥的一些地区。食物有水果、壳类种子、向日葵、绿色食物等。美丽、聪明、善解人意、长寿、品种多样化，这些特征使其备受人们喜爱。

PARROT STOLEN

Our very well loved Spectacled Amazon parrot, Pegasus, was stolen July 3rd at approximately 5:50 p.m. from Spectrum Exotic Birds where she was being boarded during our vacation. She is about 7" from tail tip to the top of her head and primarily green with white above her beak and red around her eyes and has some red coloring on her tail feathers.

Unless you have lost a pet you cannot understand our sense of loss, grief and fear for her safety and wellbeing. She had been a part of our family for five years.

If you know anyone who suddenly acquired a bird of this description after the above date, please call us at 928-5823. Leave a message if the machine is on. You may leave an anonymous message or leave your name as a reward for her recovery will be paid. Please leave enough information to allow us to investigate whether or not the bird in question is our Pegasus.

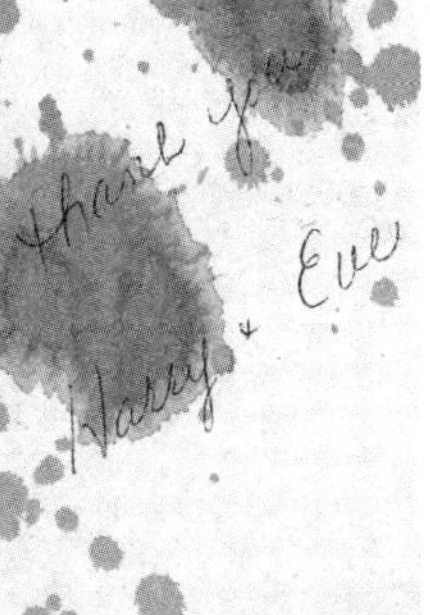

美国加利福尼亚

你快回来，我的鸟儿

它是一只小小小小鸟，披有一身可爱的黄色羽毛

及非常前卫的冲天鸟冠和害羞的绯红脸颊。

它是我丢失的鸟儿“波弟”。

请帮我找找它。

重酬答谢！

→ P180

LOST
BIRD
YELLOW
COCKATIEL
WITH RED CHEEKS
AME: "BIRDY"
LEASE CALL
REWARD

50 美元酬劳

寻鸟！

亲手驯养却抵不住出逃的诱惑，两只爱鸟终究离我而去。

它们是两只牡丹鹦鹉，于 1999 年 4 月 14 日晚上 7 点离家出走。

地址：北公园干洗店

请拨打电话联系 528-8255 约翰

或

284-5239 南希

我们会 24 小时守候！

牡丹鹦鹉小知识

牡丹鹦鹉亦称情侣鹦鹉、爱情鸟。共有 9 个品种，均产于非洲。体长一般在 15 厘米左右，体重 40~50 克。喙红色，眼及蜡膜（喙与前头部连接处的皮肤）为白色。头部黑褐色，颈部有赤黄色的环带。上胸浅绿色，背部和翼为绿色，翼端呈黑色，尾绿色，脚灰色。主要以地面上的草类种子、浆果、水果、植物嫩芽等为食。性情凶猛，以强欺弱，发情雌鸟更为突出，叫声大而杂，有时噪声扰人，还可向其他鸟进攻，这时若将雌鸟与雄鸟配对繁育则鸣声锐减，性情好转，所以饲养牡丹鹦鹉以成对为佳。

LOST "$50^{00}" Reward"

2 Hand TAMED "Love Birds"

Apr-14-99 7:00 pm

NORTH Park Cleaners

PLS CONTACT

Johnny * 528-8255

Nancy * 284-5239

* Day or Night

寻找乌鸦

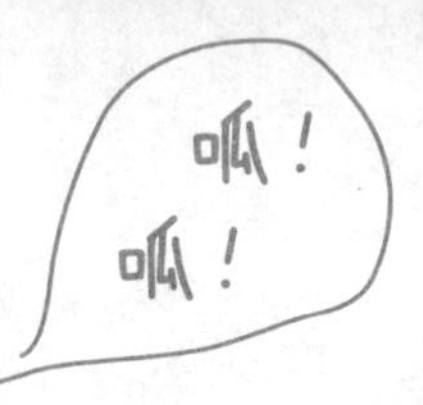

你见过我们的宠物乌鸦吗？
我们最后一次见到它在贝德福德路。

P184

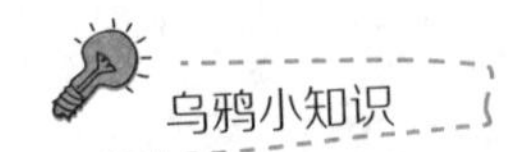

乌鸦，鸟类，嘴大而直，全身羽毛黑色，翅膀有绿光。多群居在树林中或田野间，以谷物、果实、昆虫等为食。有的地区叫老鸹、老鸦。

CAW!
CAW!
HAVE YOU SEEN OUR
PET CROW ?
LAST SEEN AT BEDFORD RD &

加拿大安大略省

Others

其他

宠物牛　仓鼠　蛇　兔子　乌龟　雪貂

悬赏！

急寻四头宠物牛。

电话（913）964-3447。

P188

REWARD

Missing —
Four head
stock cows.

Call
(913) 964-3447

美国蒙大拿州

我的宠物牛走丢了！

我的牛大概在2月16日夜（周日）至2月17日晨（周一）这个时间段走丢的。

它叫“露西安”。

如果你见到了它，请马上拨打电话27404，接它的车会立即赶过来的。

请小心疯牛病！

RIND ENTLAUFEN

Am Abend vom Sonntag, 16. Februar
auf den Montag, 17. Februar
ist uns ein Rind entlaufen.

Es hört auf den Namen Lucien und hält sich mit
Vorliebe in Restaurationsbetrieben auf.
Wenn Sie das Rind finden, bitte sofort Tel. 274'04
anrufen. Der Viehtransporter kommt unverzüglich.

VORSICHT BSE-GEFAHR

This is for a lost cow named Lucien.

小仓鼠找主人

7月5日星期六，

我在春园路上发现一只小仓鼠。

请失主见到告示联系丽萨，

电话：835-6690。

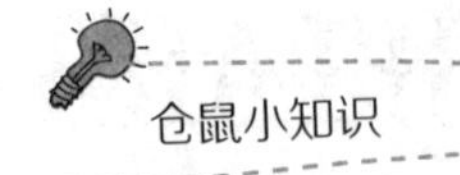

宠物鼠指人类为了观赏或趣味而饲养的鼠类，包括许多不同的物种。在现实生活中，主要以仓鼠为主。仓鼠是仓鼠亚科动物的总称，共七属十八种，主要分布于亚洲，少数分布于欧洲，其中中国有三属八种。除分布在中亚的小仓鼠外，其他种类的仓鼠两颊皆有颊囊，从臼齿侧延伸到肩部，可以用来临时储存或搬运食物回洞储藏，故名仓鼠，又称腮鼠、搬仓鼠。

Hamster Found

a hamster was found wednesday, july 5 on 'spring garden road'. owner please call:

Lisa 835-6690

寻找宠物鼠

它逃走了！它是只黑色的大个老鼠，名叫“毒药”。

求求你了，请帮我找到它。我的电话是6591137。

我好想念我的宠物鼠！

必有重谢！

P194

MISSING PET RAT

BIG BLACK RAT
ESCAPED. REWARD
TO FINDER. RATS
NAME IS POISON.
PLEASE HEIP! CAL
YAEL @ 6591137.

寻找爱蛇

颜色黑白相间，长 4 英尺。

无毒，也不咬人，见到它请不要害怕。

烦请好心人帮忙，我定会给予答谢。

蛇是无足的爬行动物的总称。正如所有爬行类一样，蛇类全身布满鳞片。所有蛇类都是肉食性动物。目前全球共有 3000 多种蛇类。其身体细长，四肢退化，无可活动的眼睑，无耳孔，无四肢，无前肢，身体表面覆盖有鳞。部分有毒，但大多数无毒。在中国的十二生肖属相中，蛇排名第 6 位。

LOST SNAKE

BLACK AND WHITE – 4' LONG

LOST SNAKE

BELOVED CLASSROOM PET

(Not poisonous, doesn't bite!)

*** REWARD ***

LOST SNAKE

求求你帮帮我吧，
11月27日走丢了！

我大名“波”，我是王蛇。

大约体长107厘米。
我的主人一定开始思念我了，
就像我现在非常思念他一样。
我发誓我真的不是故意走开，
这只是个偶然事件。
如果善良的你在校园见到我，
请告诉我的主人维克·伊特诺
我的位置，他一定会乐意给
你报酬的。谢谢。

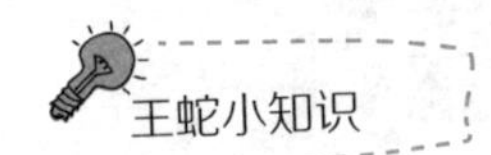

王蛇小知识

王蛇（*kingsnake*），又名皇帝蛇。王蛇属（学名：*Lampropeltis*）是蛇亚目游蛇科下的一个蛇属，属下的蛇类通称王蛇。当中包括著名的牛奶蛇。分布于加拿大东南部至厄瓜多尔。成蛇体长一般1~1.5米，但有些长达2.1米。王蛇是无毒的大型游蛇，取食种类广泛，包括小型哺乳类、鸟类、蛇类、蜥蜴、两栖类和鸟蛋。王蛇之所以叫王蛇，必有其过蛇之处，其自身本无毒，但却以毒蛇为食，原因就是它对毒蛇的毒性几乎免疫，在原产地经常以响尾蛇或铜斑蛇为食。王蛇一般都是以蟒蛇的缠绕方式使猎物窒息死亡后再吞食。如果有人住家区域有鼠类为患，只要养条王蛇，保证方圆几公里之内鼠类绝迹。

HELP, I'M LOST! 11/27

← 3½ FT →

My name is "BO" & I am a king snake. My owner surely misses me as much as I miss him. I swear I didn't mean to get lost, it just sorta happened. If you see me around campus, please call my owner Vick Itone & tell him where I'm located. I am sure my owner will gladly provide a reward. Thanks, Bo.

美国加利福尼亚

发现家养的小兔子

[illegible]年7月6日

它披着褐色毛发，黑色小鼻子。

请主人前来认领或有责任心的朋友也可以

前来领养，给它一个温暖的家。

电话：696-1087

电话：696-1087

电话：696-1087

电话：696-1087

电话：696-1087

→ P200

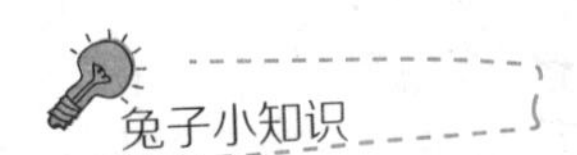

兔子，哺乳纲、兔形目全体动物的统称。短尾，长耳，头部略像鼠，上嘴唇中间裂开，尾短而向上翘，后腿比前腿稍长，善于跳跃，跑得很快，并且可爱机灵。有家养的和野生的。

Found: Domestic

6/7/94

(Rabbit)

- Tan with dark nose

- Call to reclaim <u>or</u>
- Available for adoption to <u>responsible</u>, <u>permanent</u> home

RECLAIM OR ADOPT
RABBIT 696-1087

RECLAIM OR ...
RABBIT 696-1087

RABB...

RECLAIM OR ADOPT
RABBIT 696-1087

...CLAIM OR ADOPT
...ABBIT 696-1087

RECLAIM OR ADOPT
RABBIT 696-1087

RECLAIM OR ADOPT
RABBIT 696-1087

RECLAIM OR ADOPT
RABBIT 696-1087

如果你发现一只像这样带有棕色斑点的兔子，请致电：

474543

（无论“生”或“死”！）

Si vous trouve
un lapin comme ça
avec les tache
brun appelez

474573

S.V.P.

(vivant ou mort)

Alive or dead.

法国

如果你丢失了一只小龟
或是知道谁弄丢了小龟，
请打电话 864-4792
和我联系。

龟小知识

龟，俗称乌龟，泛指龟鳖目的所有成员，是现存最古老的爬行动物。特征为身上长有非常坚固的甲壳，受袭击时龟可以把头、尾及四肢缩回龟壳内。大多数龟均为肉食性，以蠕虫、螺类、虾及小鱼等为食，亦食植物的茎叶。龟通常可以在陆上及水中生活，亦有长时间在海中生活的海龟。龟亦是长寿的动物，自然环境中有超过百年寿命的。

IF YOU LOST A
TURTLE OR KNOW
SOMEONE WHO
DID:
CALL 864-41796

美国加利福尼亚

寻找小龟

请帮忙找到他，拜托了！

GRANT大街1620号

电话：294-9805

P206

TURTLE

FIND HIM

1620 GRANT

294 - 9805

美国科罗拉多州

丢失的小雪貂现已找到
现寻爱心人士收养

我们住在司帕蒂娜西侧，拿骚和鲍德温之间。

由于我们养了两只猫，所以无法继续养小貂了。

如有对喂养这只小雪貂感兴趣的朋友，

请关注“人文社区”。联系电话：

332-2273

P208

（附：小雪貂是在1995年8月26日周六下午

被送回的）

貂小知识

貂，哺乳动物，身体细长，四肢短，耳朵三角形，听觉敏锐，种类较多，有石貂、紫貂等。大部分貂属动物都居住在树上，以松鼠为食，它们的食物还包括鸟和蛋。貂在中国主要产于东北地区，有多个品种。属珍贵毛皮动物。

ON WEST SIDE OF SPADINA
BETWEEN NASSAU & BALDWIN

WE HAVE 2 CATS AND CANNOT
KEEP THE FERRET FOR LONG,
SO CHECK WITH THE "HUMANE
SOCIETY" (392-2273) AS WELL.

(FERRET FOUND SATURDAY (P.M.) 26 AUGUST '95)

请帮我们找到我们的好朋友

小貂儿，你在哪里？

他叫“波波”，也叫“波拜”。

重金奖励！

我们于2月21日晚上8点半发现他离家出走，

自此之后他便杳无音信。

我们怀疑他是从我家前门逃走了，我家住

在11号大街，如果你见到他，请帮忙抓住他。

他很友好，从不咬人。灰色皮毛，一只耳，

年方一岁。

PLEASE HELP US FIND OUR FRIEND

LOST FERRET

"Boo-boo" or "Boo-bear"

REWARD

Call also, if you have heard about a newly "found" ferret. This is how animals get far away from home. They are found and taken to a new location.
Sable-brown tattoo dots in one ear, male, neutered, one year old

LAST SEEN @ HOME FEB 21ST 8³⁰ pm.
MAY HAVE ESCAPED VIA THE FRONT DOOR
OF 14th AVE. NE. IF YOU SEE HIM
PLEASE TRY TO CAPTURE HIM. PEOPLE-
FRIENDLY. WILL NOT BITE.

我的爱宠走失了！

她名叫"莉莉"。
她是一只黑白相间的雪貂，
如果你见到她，
请尽快联系我，谢谢帮忙。

电话：781-032

P212

LOST: Ferret

COLOR: "BLACK" "AND" "WHITE"

NAME "LiL"

"CALL... 781-032

我弄丢了我的三明治

我的热乎乎、香喷喷的三明治竟然被偷了，我在克里小镇第五大街的科斯莫熟食店里买下它，然后它不翼而飞了。这种三明治是我最喜欢的了，好想找到它。请叔叔阿姨帮帮我这个可怜的小孩子吧。

必重谢！

P214

MISSING

GRILLED CHEESE!

IF YOU KNOW WHO STOLE THIS SANDWICH:

last seen:
Kosmo Deli
in Kerrytown
on Fifth Ave.

very friendly,
loves kids.

PLEASE CONTACT

REWARD

DESPERATELY SEEKING SANDWICH. PLEASE HELP

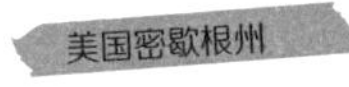

后　　记

寻找宠物海报小贴士

每年仅在美国就有数以万计的宠物走丢。宠物找回的几率很低，百分之九十以上走丢的猫狗都没能找回。张贴寻找宠物海报是找回宠物的最好方法。

标题：通常写法如“寻猫启事”，容易识别。

描述：写清楚宠物的基本信息，包括品种、毛发颜色、眼睛颜色、体形大小、性别、年龄、特征等等。记得保留个别信息用来识别那些为了索取报酬而谎称找到宠物的人。

名字：有些人建议不要在海报上泄露宠物的名字，因为有些人可能会借此取得宠物的信任。

宠物标识：项圈、标牌、微芯片、纹身等。

时间和地点：宠物丢失的地点及经过。

联系方式：你的名字和电话，最好不要写家庭住址。

照片：在海报上附上你家宠物的照片，没有的话类似的画像也可，最好是全身像。

报酬：奖励会激励人们花费时间检查他们的车库或库房，但

是不要列出报酬的最高数额，因为可能会降低感召力。如果回报数额巨大，很可能会陷入电话诈骗。

用影印机制作海报效果好且价钱低廉，当然如果你愿意多花钱做彩印，效果会更好。不要用喷墨打印机做海报，因为若是碰到雨雪天气的话这种海报会变得模糊一片。

你贴出的海报越多，你寻回宠物的机会就越大。电线杆是张贴海报的最佳位置，尽管一些市政当局并不允许在这种公共场所张贴海报和传单。不要将海报订在树上，尝试将其贴在宠物经常出现的社区里：杂货店、动物医院、宠物店、图书馆、自助洗衣店、便利店、操场及人多地段。贴海报的时候，留意观察有没有贴宠物招领的海报。将海报贴在可以平视的高度，让过路的人和驾车的人都方便阅读。经常回去检查海报是否还在原处，因为很有可能会出于各种原因被人移除。经常用新的海报替换由于日久而残破不清的海报。

如果有人来电话说在某地见过丢失的宠物，找出他电话中提供的所在位置并抓紧时间赶到那个区域张贴大量海报。如果有人来电说找到了你的宠物，你要问他问题来核实他所说是否属实。不要在电话里谈及太多细节问题，谨防那些诈取报酬的人。如果你找到了宠物，记得把自己张贴过的海报全揭下来。

41% 的美国人都在家里存有自己宠物的照片

17% 的人将宠物的照片放入自己的钱包中

1300 万只猫曾经过生日

16% 的狗可以跟主人同床睡觉

美国公犬和母犬的数量几乎一样多

20% 的宠物狗住在专门搭建的狗窝里

每降生一个人，便会有 7 只小狗和小猫同时出生

28539216 个宠物主人为他们的宠物狗买圣诞礼物

95% 的人每天跟自己的狗拥抱

最好的看门犬：罗威纳犬

最差的看门犬：侦探猎犬

每一只母猫和她的后代可以在 7 年内产下一共 42 万只猫

每年美国人在宠物上的消费为 285 亿美元

在美国，20% 的宠物主人爱狗胜过最好的朋友

6% 的人对自己宠物狗的依恋更甚于自己的配偶

83% 的宠物主人需要宠物的陪伴

最高生产纪录：1944 年 6 月 19 日降生了 23 条美国猎狐犬

在美国，最常用的宠物名字是马克斯，随后依次是山姆、女士、思慕克、大熊

／鸣　　　　谢

特别感谢多年来为本项收集贡献海报的每一个人。尤其感谢Sonja Ahlers, George Banton, Peter Buchanan-Smith, Patricia Collins, Rachel Crossley, Mike Dyar, Ronald Consalves, Grant Heaps, Derek McCormack, Mark Pawson, Julee Peaslee, Stangroom, Gerardo Yepiz.

LOST: LOST AND FOUND PET POSTERS FROM AROUND THE WORLD by IAN PHILLIPS

图书在版编目（CIP）数据

喵了个咪啊去哪儿了？/【加】伊恩·菲利普斯（Phillips, I.）编著；夏楠 译.—合肥：黄山书社，2014.6
ISBN 978-7-5461-4600-3

Ⅰ.①喵… Ⅱ.①伊… ②夏… Ⅲ.①宠物－图集 Ⅳ.①S865.3-64

中国版本图书馆CIP数据核字（2014）第115161号

版权合同登记号：12121190

喵了个咪啊去哪儿了？
MIAO LE GE MI A QU NAER LE
【加】伊恩·菲利普斯 编著 夏楠 译

出版人 任耕耘
策划 任耕耘 杨雯
责任编辑 汪盎然
特约编辑 赵迪秋
装帧设计 齐娜

出版发行 时代出版传媒股份有限公司（http://www.press-mart.com）
黄山书社（http://www.hspress.cn）
官方直营书店（http://www.hsssbook.taobao.com）
营销部电话：0551-63533762 63533768
（合肥市政务文化新区翡翠路1118号出版传媒广场7层 邮编：230071）
经销 新华书店
印刷 安徽新华印刷股份有限公司

开本 880×1230 1/32 印张 7.25 字数 100千字
版次 2014年6月第1版 印次 2014年6月第1次印刷
书号 ISBN 978-7-5461-4600-3 定价 28.00元